I0066524

Environmental Waste Management

Environmental Waste Management

Edited by **Victor Bonn**

New York

Published by Syrawood Publishing House,
750 Third Avenue, 9th Floor,
New York, NY 10017, USA
www.syrawoodpublishinghouse.com

Environmental Waste Management
Edited by Victor Bonn

International Standard Book Number: 978-1-68286-085-4 (Hardback)

Printed in the United States of America.

Contents

Preface

Waste management is an emerging branch of environmental sciences dealing with the treatment and disposal of waste. This discipline deals with the clearance, treatment and management of both environmental and industrial waste. This book gives an elucidative account of the different techniques and practices involved in management of solid wastes with the help of biomolecular tools and traces. It compiles theoretical and analytical approaches to reduce the potential waste disposal risks in both developed and developing countries. This book will be beneficial to students and researchers who are involved in resolving various scientific and technical issues related to this field.

This book is a comprehensive compilation of works of different researchers from varied parts of the world. It includes valuable experiences of the researchers with the sole objective of providing the readers (learners) with a proper knowledge of the concerned field. This book will be beneficial in evoking inspiration and enhancing the knowledge of the interested readers.

In the end, I would like to extend my heartiest thanks to the authors who worked with great determination on their chapters. I also appreciate the publisher's support in the course of the book. I would also like to deeply acknowledge my family who stood by me as a source of inspiration during the project.

Editor

1

On the Durability of Nuclear Waste Forms from the Perspective of Long-Term Geologic Repository Performance

Yifeng Wang , Carlos F. Jove-Colon, Robert J. Finch

Sandia National Laboratories, P. O. Box 5800, Albuquerque, NM 87185, USA

Abstract: High solid/water ratios and slow water percolation cause the water in a repository to quickly (on a repository time scale) reach radionuclide solubility controlled by the equilibrium with alteration products; the total release of radionuclides then becomes insensitive to the dissolution rates of primary waste forms. It is therefore suggested that future waste form development be focused on conditioning waste forms or repository environments to minimize radionuclide solubility, rather than on marginally improving the durability of primary waste forms.

Key words: nuclear waste management; waste form; backfill; chemical control; deep geologic reposi tory; radionuclide solubility

---

1. Introduction

Durability is commonly considered to be the most important attribute of nuclear waste forms (WF) [1,2,3]. A great deal of effort has been devoted to creating durable waste forms, ranging from borosilicate glasses to crystalline mineral analogues [4,5], with the assumption that a slow waste form dissolution rate is the most effective means of limiting the release of radionuclides (RN) from a repository to the human accessible environment. This assumption is based on an untested premise that the kinetic dissolution of a primary waste form is the rate-limiting step for radionuclide release. In this short communication, we demonstrate that this may not be the case in an actual repository environment. Once in contact with incoming water, a primary waste form in a repository will eventually degrade into various more stable alteration products (or secondary waste forms). We show that the long-term rate at which radionuclides are released from a repository will be controlled by the solubilities of the alteration products rather than by the dissolution rates of the primary waste forms. Therefore, current requirements for waste form development need to be re-evaluated.

2. Results

Long-term performance assessments (PA) of a geologic repository generally model the degradation of a waste form as a kinetic process, whereas alteration products are assumed to be in chemical equilibrium with the contacting water [6]. Such treatment is referred to as the partial equilibrium approach, which assumes no supersaturation in the system. This approach has been invoked by the IAEA [7] and has been used to assess long-term waste isolation in a geologic repository [8,9]. This approach seems appropriate, if we look at the dissolution and alteration of natural materials in subsurface environments. We have calculated mineral-saturation indices for groundwater samples collected from Columbia Basin [10], which indicate that a majority of water samples are close to equilibrium with calcite and cristobalite, with little or no oversaturation with other low temperature minerals. This is probably due to the presence of a large number of nucleation and growth sites on mineral surfaces in a rock-weathering system, which effectively lowers the saturation degree for secondary mineral precipitation. As shown in Figure 1, calcium release due to basalt weathering can be very well predicted (within one log unit) by assuming calcite to be the solubility-controlling mineral for dissolved calcium. A similar observation can be made on uranium ore deposits. For example, Lu et al. [11] describes the oxidation of uraninite from a uranium ore deposit in China and the formation of secondary solubility-controlling uranyl-bearing solids.

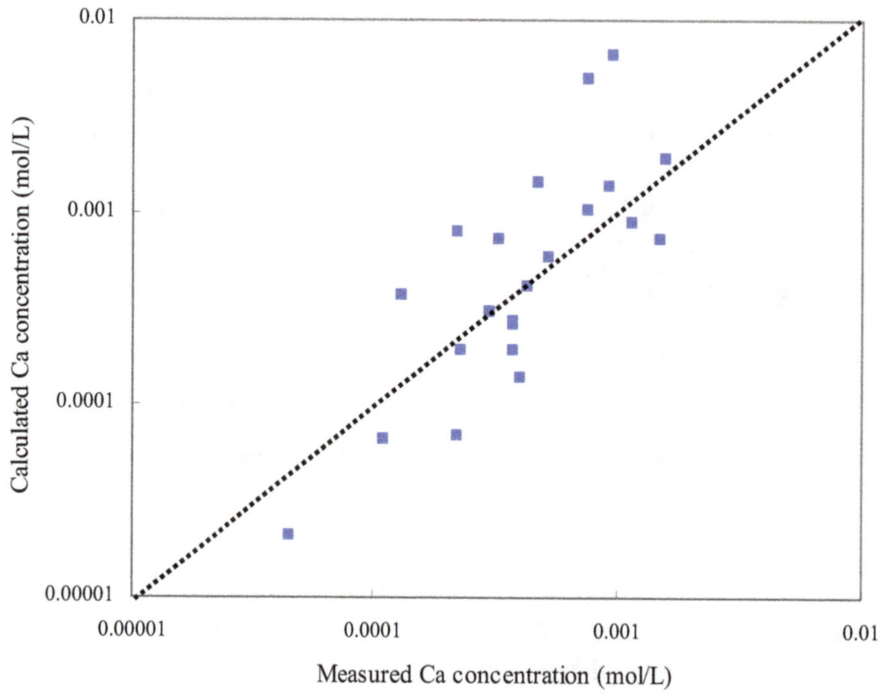

Figure 1. Dissolved calcium concentrations predicted with the assumption that calcium release is controlled by equilibrium with calcite. The good agreement between the calculated and the measured concentrations justifies the partial equilibrium assumption generally used in modeling of waste form degradation in a deep geologic repository.

Based on the assumption of partial equilibrium, in an actual repository environment, waste degradation and radionuclide release can be divided into three stages (Figure 2): Stage I, in which the dissolved concentration of the radionuclide has not reached its solubility limit; Stage II, in which the radionuclide concentration has reached its solubility limit and continues to maintain equilibrium with a solubility-controlling secondary phase; and Stage III, in which the dissolved radionuclide concentration drops rapidly to zero due to the disappearance of both the primary waste form and the solubility-controlling secondary phase. The actual release mode of a radionuclide depends on how fast the repository system can reach Stage II, which is controlled by three factors: (1) the rate of radionuclide release from the waste form, (2) the rate at which water flows through the repository, and (3) the radionuclide's solubility under repository conditions.

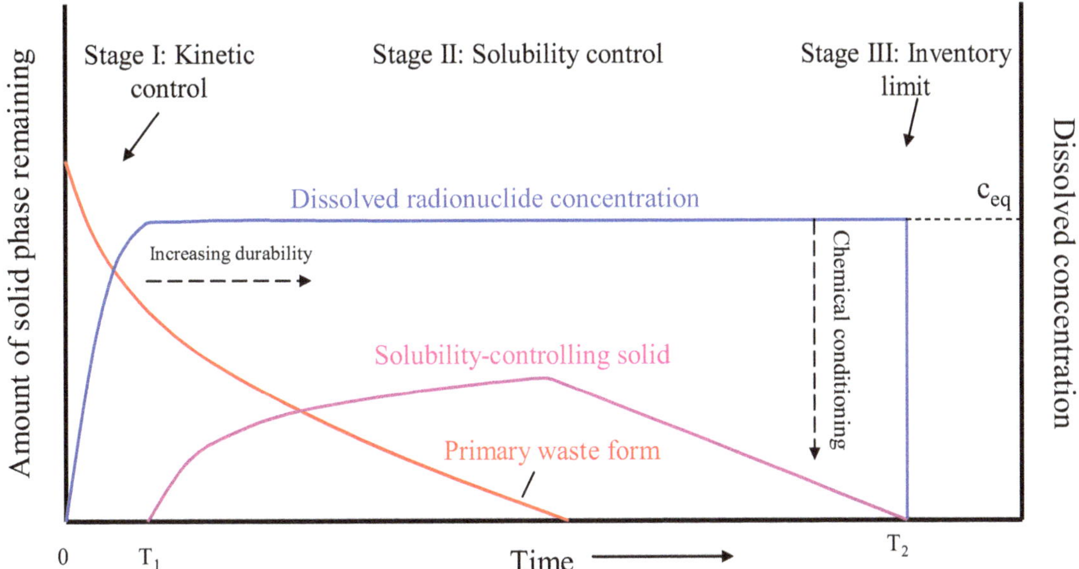

Figure 2. Schematic diagram of waste degradation and radionuclide release in a repository environment under the partial equilibrium assumption. The duration of stage I is generally much shorter than that of stage II. Stage III may never be reached for a radionuclide with a relatively large initial inventory and a slow decay rate. As demonstrated in this paper, chemical conditioning can be more effective in reducing the total radionuclide release than further improving waste form durability.

In Stage I, the concentration of a specific radionuclide in a disposal system can be modeled as follows:

$$V_T\phi\frac{dc}{dt} = -\frac{\alpha}{m_w}\frac{dM}{dt} - vc \qquad \text{for} \quad c < c_{eq} \qquad (1)$$

$$\frac{dM}{dt} = -M \cdot A \cdot R \qquad (2)$$

where V_T is the total volume a disposal system (dm^3); ϕ is the porosity in the disposal system after waste emplacement; c is the dissolved concentration of a radionuclide of interest ($mol \cdot L^{-1}$); t is the time since disposal (s); α is the waste loading factor ($g_{RN} \cdot g_{WF}^{-1}$); m_w is the molecular weight of the radionuclide ($g_{RN} \cdot mol^{-1}$); M is the total mass of a waste form in the disposal system (g_{WF}); v is the rate of water flow percolating through the disposal system ($L \cdot s^{-1}$); c_{eq} is the solubility of the radionuclide (mol/L); A is the specific surface area of waste form ($m^2 \cdot g_{WF}^{-1}$); and R is the surface-normalized dissolution rate of the waste form ($g_{WF} \cdot m^{-2} \cdot s^{-1}$). R depends on various factors such as the chemical affinity for waste form dissolution and the coating of alteration products. For simplicity, we assume that the effect of those factors is captured by wide ranges of variability in dissolution rates reported in the literature. In Equation (1), we ignore the concentration change due to radioactive decay. This simplification does not change the overall conclusion drawn in this paper, especially for long-lived radionuclides in which we are interested. Also here we use term "radionuclide" interchangeably with "radioelement" when we refer to "radionuclide solubility".

Solving Equation (2) and plugging the solution into Equation (1), we obtain:

$$V_T\phi\frac{dc}{dt} = -\frac{\alpha ARM_0}{m_w}e^{-A\cdot R\cdot t} - vc \quad (3)$$

The initial mass of waste form M_0 is $M_0 = \rho(1-\phi)V_T$, where ρ is the density of waste form ($g_{WF}\cdot m^{-3}$). Equation (3) can be scaled into:

$$\frac{du}{d\tau} = e^{-\beta\tau} - \gamma u \qquad \text{for } u < 1 \quad (4)$$

with

$$\tau = \frac{t}{T} \qquad T = \frac{c_{eq}m_w\phi}{\alpha AR\rho(1-\phi)} \qquad \beta = \frac{c_{eq}m_w\phi}{\alpha\rho(1-\phi)}$$

$$u = \frac{c}{c_{eq}} \qquad \gamma = \frac{c_{eq}m_w\phi}{\alpha AR\rho(1-\phi)}\frac{1}{T_r} \qquad T_r = \frac{V_T\phi}{v} \quad (5)$$

Parameter T represents a typical time scale for the variation of dissolved radionuclide concentration, over which $du/d\tau$ is on the order of 1; T_r characterizes the residence time water inside the disposal room. For typical parameter values given in Table 1, $\beta < 10^{-4}$, and thus Equation (4) can be simplified to:

$$\frac{du}{d\tau} \approx 1 - \gamma u \qquad \text{for } u < 1. \quad (6)$$

The small value of β is a characteristic of subsurface systems with high solid/water ratios [12].

Table 1. Typical model parameter values

Parameter	Values	Comment
Radionuclide solubility (c_{eq})	10^{-5} mol/L	
Molecular weight (m_w)	237 g/mol	Use Np as an example
Porosity (ϕ)	0.4	
Density of waste form	2.7–2.9 g/cm^3	Use high-level waste glass as an example.
Reactive surface area of waste form	0.01 m^2/g	
Pore volume in a WIPP waste panel	10^3–10^4 m^3	[14]

Maximum rate of water flow pore upper boreholes in a WIPP waste panel	0.55 m^3/yr	[14]
Volume of a waste container in Yucca Mountain repository	15 m^3	[13]
Rate of water flow percolating through a Yucca Mountain waste container	0.1–100 L/yr	[13]
Water flow rates for other repository-relevant formations	Approx. 10^{-5} L/yr for clay (Meuse/Haute-Marne site); 100–10,000 L/yr for granite	[28]

For a solubility control of radionuclide release, it is required that $\gamma < 1$; that is,

$$T_r > \frac{c_{eq} m_w \phi}{\alpha A R \rho (1-\phi)}. \quad (7)$$

This ensures that the rate of radionuclide dissolution into the solution exceeds the rate of radionuclide transport away from the system by water advection. By solving Equation (6) and then setting u = 1, the time for a dissolved radionuclide to reach its solubility limit (T_l) can be calculated as follows:

$$T_1 = -\frac{T}{\gamma}\ln(1-\gamma) = -T_r \ln\left[1 - \frac{c_{eq} m_w \phi}{\alpha A R (1-\phi)\rho T_r}\right] \quad (8)$$

As indicated in Figure 3, with increasing the residence time of water in the disposal system or increasing dissolution rate of the primary waste form, the concentration of the radionuclide will change from being kinetically controlled to being solubility-controlled. The residence time depends on the actual water flow rate and the total pore volume of the disposal system. An actual disposal system can be a waste container or a waste panel. In the Yucca Mountain repository, radioactive wastes will be container in steel/alloy packages, and each waste package, with a total internal volume of ~15 m^3 9 can be considered as a separate disposal system (note that ignoring the WP and assuming the tunnel immediately surrounding the waste effectively increases the value of pore volume and the corresponding value of T_r). Choosing a typical flow rate in the range 0.1–100 L/year/waste package [13], we estimate the minimum water residence time to be ~60 years. Similarly, for the Waste Isolation Pilot Plant (WIPP), a disposal system can be defined as an individual waste panel, each with the pore volume of 0.3–1.0x10^4 m^3 (Helton et al., 1998). The maximum rate of water flow through boreholes during a human intrusion scenario is estimated to be ~5.5 m^3/year [14]. The minimum water residence time in the WIPP is thus estimated to be ~5,000 year. Therefore, as shown in Figure 3, in an actual repository environment, even for the most durable waste forms, the radionuclide release from the near-field of a repository will likely be controlled by the solubility limit of the radionuclide.

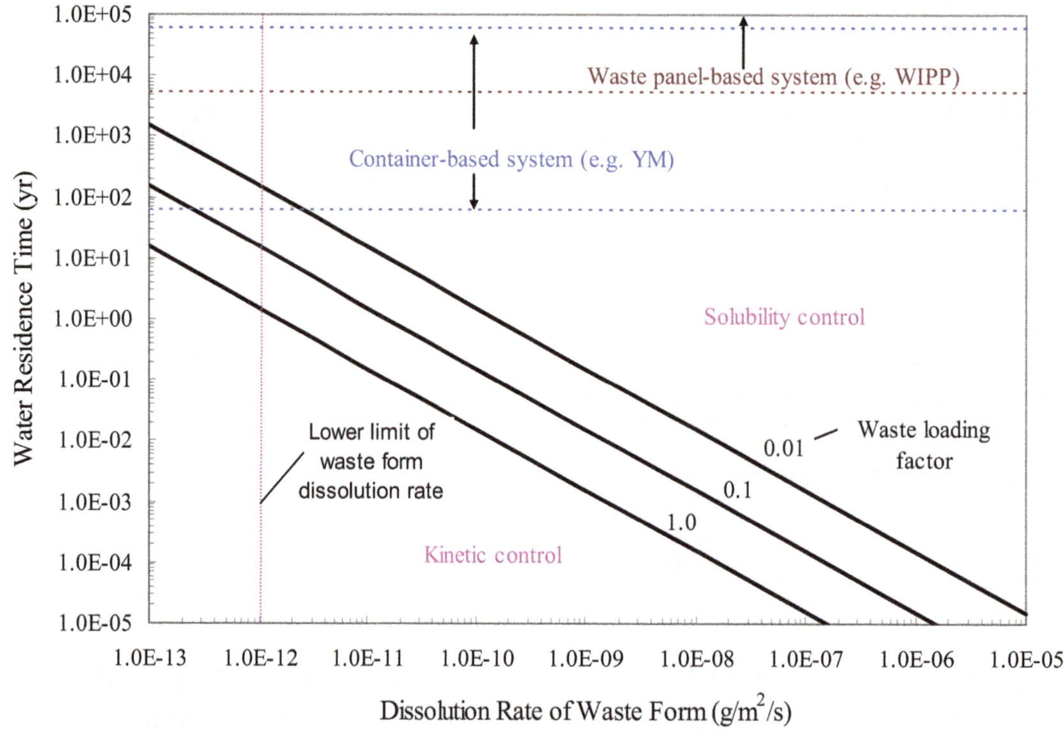

Figure 3. Radionuclide release modes as controlled by waste form dissolution rate and waste loading factor as well as by water residence time. It is apparent that in an actual repository environment radionuclide release is most likely controlled by solubility of alteration phases. The lower limit of waste form dissolution rate is assumed to be the minimum dissolution rate of zirconia (see Figure 4). The parameter values used in constructing this diagram are shown in Table 1.

For a given repository environment, the time for a radionuclide to reach its solubility limit is determined by the dissolution rate of the waste form that hosts the radionuclide. We have compiled dissolution rates for various waste forms. As shown in Figure 4, for all waste forms considered, the time for a radionuclide to reach its solubility limit is generally less than 300 years. Compared with a typical repository regulatory time (T_{reg}) (10^4–10^6 years), this transient time is negligible. Therefore, the accumulative release of a specific radionuclide can be approximated by:

$$\text{Total release} = \int_0^{T_{reg}} c \cdot v dt \approx \int_{T_1}^{\min(T_2, T_{reg})} c_{eq} v dt \ . \tag{9}$$

As indicated in Equation (9), there are two ways to reduce the total radionuclide release. One way is to reduce the water flow rate v with an engineered physical barrier, for example, encapsulate a waste form with a low-permeability clay layer [15] (see Table 1). The problem with an engineered barrier is that it is difficult to demonstrate the long-term integrity of barrier over a regulatory time period. The other way, which we believe will be more effective in reducing radionuclide release, is to chemically condition waste forms or repository environments to minimize radionuclide solubility.

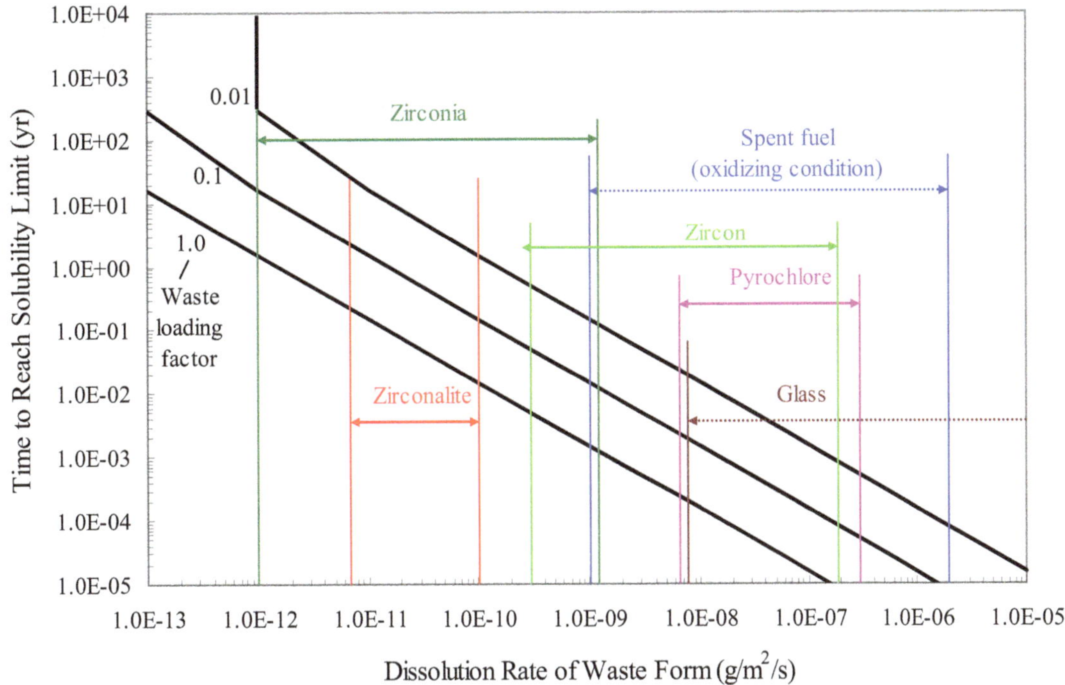

Figure 4. Time for a radionuclide to reach its solubility limit as a function of waste form dissolution rate. The water residence time is assumed to be 200 years. Dissolution rates are taken from [22] for glass, [23] for spent fuel, [24,25] for pyrochlore, [26] for zirconalite, [5] for zirconia, and [2,27] for zircon. The parameter values used in constructing this diagram are shown in Table 1.

The concept of conditioning repository environments can be best demonstrated with the Waste Isolation Pilot Plant (WIPP) [16], which is located in a salt bed in southern New Mexico and designed by U.S. Department of Energy for permanent disposal of defense-related transuranic wastes. WIPP wastes contain a large quantity of organic materials and various nutrients. Thus, there is a concern about the potential impact of microbial CO_2 generation on actinide solubilities. To mitigate this effect, a sufficient amount of MgO will be added to the repository with the backfill. Hydrated MgO will react with CO_2 to form magnesite:

$$Mg(OH)_2 + CO_2 \rightarrow MgCO_3 + H_2O. \quad (10)$$

Reaction (10) will buffer CO_2 fugacity at ~ 10^{-7} atm. This low CO_2 fugacity implies that Reaction (10) will remove practically all CO_2 from both gaseous and liquid phases in the repository. A thermodynamic equilibrium calculation using the EQ3/6 code shows that the addition of MgO will buffer pH around 10 for WIPP brines [16]. Under these chemical conditions, actinide solubilities become minimal (Figure 5).

As shown in Figure 5, by appropriately conditioning the near-field environments the dissolved concentration of a radionuclide can be reduced by orders of magnitude. The chemical composition of a waste form can strongly affect the near-field chemistry of a repository, and so an appropriate choice of waste form composition can be an important aspect of overall chemical conditioning. In this sense, amorphous materials may have an advantage over crystalline materials, although the former are generally less "durable" than the latter. Amorphous materials such as glasses have considerable flexibility in incorporating various chemical components. It is known that many radionuclides, especially actinides, can form sparingly soluble phosphates within the glass corrosion layers [17,18] or as alteration products in uranium deposits [11]. Phosphate has also been used for immobilization of heavy metals in soils and sediments [19,20]. Therefore, it may be desirable to formulate a high-level waste (HLW) glass by adjusting its phosphate content or even to employ a phosphate glass [21] to minimize radionuclide solubility during glass degradation.

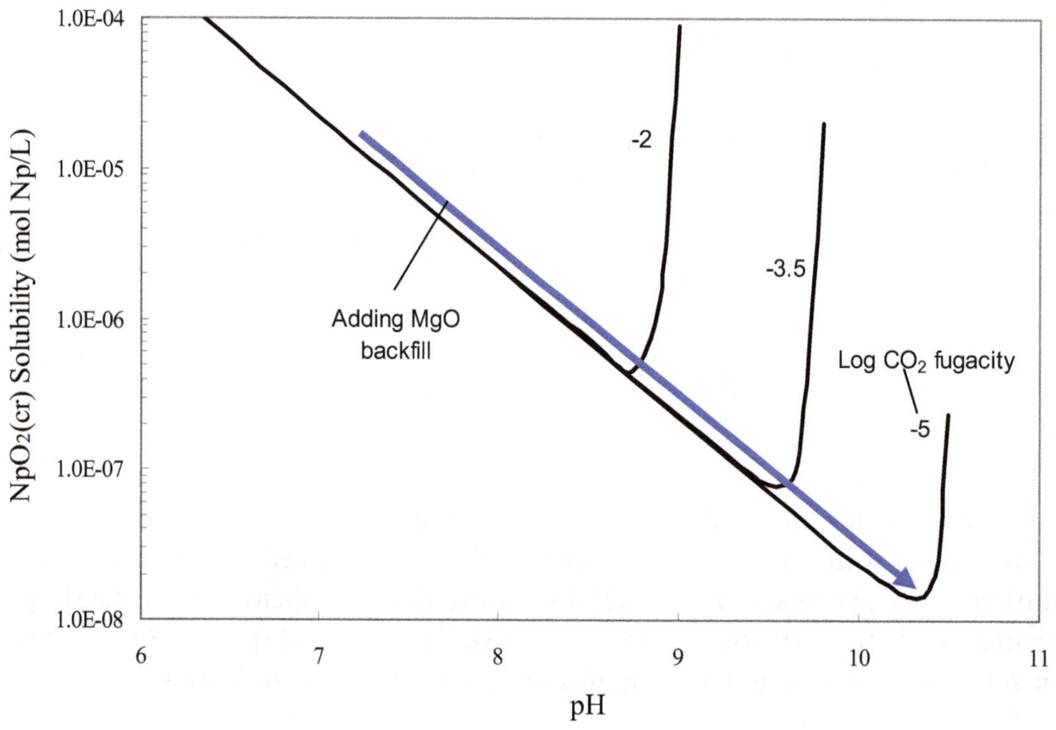

Figure 5. NpO2(cr) solubility as a function of pH and partial pressure of CO_2. Addition of MgO to the repository greatly reduces Np solubility.

For a given repository environment, dissolved concentrations of some radionuclides may never become solubility-limited, and their releases are generally limited by the dissolution rate of the primary waste form. Clearly, long-term durability of a waste form — especially one that contains a substantial inventory of such potentially mobile radionuclides — is a crucial aspect in limiting radionuclide releases; we are not suggesting otherwise. On the other hand, the proposed chemical conditioning concept may allow us to design waste forms or backfill materials that include chemical components that can precipitate low-solubility radionuclide-bearing solids that would otherwise not be stable. For example, iodine-129, which is considered a highly mobile radionuclide in nearly all repository environments, can potentially form insoluble solids in either a reducing or an oxidizing environment, provided that appropriate chemical components (such as Cu^+) are added to the host waste form or backfill materials.

Radiation damage has been a concern for the long-term performance of a waste form [2]. The existing studies in this area have been exclusively focused on primary waste forms [29,30,31]. Our analysis, however, suggests that the future focus of these studies be shifted to evaluating the potential effect of radiation damage on the stabilities of secondary mineral phases that will directly control dissolved radionuclide concentrations.

3. Conclusion

In summary, due to high solid/water ratios and slow groundwater percolation rates, the concentrations of radionuclides dissolved in water flowing through a repository are expected to reach solubility control by radionuclide-bearing alteration products well within typical regulatory timeframes. Consequently, total release of radionuclides will become insensitive to the dissolution rates of primary waste forms. Our analysis suggests that future waste-form development should be increasingly focused on conditioning waste forms or repository environments to minimize radionuclide solubility, rather than on striving for marginal improvements to the durability of primary waste forms.

Acknowledgment

Sandia National Laboratories is a multi-program laboratory managed and operated by Sandia Corporation, a wholly owned subsidiary of Lockheed Martin Corporation, for the U.S. Department of Energy's National Nuclear Security Administration under contract DE-AC04-94AL85000.

Conflict of Interest

All authors declare no conflicts of interest in this paper.

References

1. Ringwood AE, Kesson SE, Revve KD, et al. (1988) Synroc (for radiowaste solidification), in *Radioactive Waste Forms for the Future*, W. Lutze and R. C. Ewing (eds), North-Holland, Amsterdam, 233-334.
2. Ewing RC (1999) Nuclear waste forms for actinides. *Proc Natl Acad Sci* 96: 3432-3439.
3. Peters MT, Ewing RC (2007) A science-based approach to understanding waste form durability in open and closed nuclear fuel cycles. *J Nucl Mater* 362: 395-401.
4. Grambow B (2006) Nuclear waste glasses – how durable? *Elements* 2: 357-364.
5. Lumpkin GR (2006) Ceramic Waste Forms for Actinides *Elements* 2: 365-372.
6. SNL (Sandia National Laboratories) (2008) Total System Performance Assessment Model/Analysis for the License Application, Sandia National Laboratories, Albuquerque, New Mexico.
7. IAEA (2003) Technical Reports Series No. 4I3, Scientific and Technical Basis for the Geological Disposal of Radioactive Wastes.
8. Murphy WM (2004) Measures of geologic isolation. *Mat Res Soc Symp Proc* 824: cc3.5.1-cc3.5.9.
9. Kienzler B, Metz V, Lutzenkirchen J, et al. (2007) Geochemically based safety assessment. *J Nucl Sci Technol* 44: 470-476.
10. Hearn PP, Steinkampf WC, Bortleson GC, et al. (1985) Geochemical Controls on Dissolved Sodium in Basalt Aquifers of the Columbia Plateau, Washington. Water-Resources Investigations Report 84-4304. Tacoma, Washington: U.S. Geological Survey.
11. Lu L, Chen F, Ewing RC, et al. (2007) Trace element immobilization by uranyl minerals in granite-hosted uranium ores: Evidences from the Xiazhuang ore field of Guangdong province, China. *Radiochim Acta* 95: 25-32.
12. Ortoleva P, Merino E, Moore C, et al. (1987) Geochemical self-organization, I. Feedbacks, quantitative modeling. *Am J Sci* 287: 979–1007.

13. Wang Y, Jove-Colon CF, Mattie PD, et al. (2010), Thermal, hydrodynamic, and chemical constraints on water availability inside breached waste packages in the Yucca mountain repository. *Nucl Technol* 171: 201-219.
14. Helton JC, Bean JE, Berglund JW, et al. (1998) Uncertainty and Sensitivity Analysis Results Obtained in the 1996 Performance Assessment for the Waste Isolation Pilot Plant. Sandia National Laboratories, Albuquerque, New Mexico.
15. ANDRA (2005) Dossier 2005 Argile: Evaluation of the feasibility of a geological repository in an argillaceous formation.
16. Wang Y, Brush LH, Bynum RV (1997) Use of MgO to mitigate the effect of microbial CO_2 production in the Waste Isolation Pilot Plant. WM'97, March 2 - 6, 1997, Tucson, Arizona..
17. Bates JK, Bradley JP, Teetsov A, et al. (1992) Colloid formation during waste form reaction: Implications for nuclear waste disposal. *Science* 256: 649-651.
18. Buck EC, Bates JK (1999) Microanalysis of colloids and suspended particles from nuclear waste glass alteration. *Appl Geochem* 14: 635-653.
19. S. B. Chen, Y. G. Zhu, Y. B. Ma (2006) The effect of grain size of rock phosphate amendment on metal immobilization in contaminated soils. *J. Hazardous Mater* 134: 74-79.
20. Ndiba P, Axe L, Boonfueng T (2008) Heavy Metal immobilization through phosphate and thermal treatment of dredged sediments. *Environ. Sci Technol* 42: 920-926.
21. Kim CW, Day D (2003) Immobilization of Hanford LAW in iron phosphate glasses. *J Non-Crystalline Solids* 331: 20-31.
22. Ojovan MI, Hand RJ, Ojovan NV, et al. (2005) Corrosion of alkali-boro silicate waste glass K-26 in nono-saturated conditions. *J Nucl Mater* 340: 12-24.
23. BSC (Bechtel SAIC Company) (2004) CSNF Waste Form Degradation: Summary Abstraction. ANL-EBS-MD-000015 REV 02. Las Vegas, Nevada: Bechtel SAIC Company.
24. Strachan DM, Scheele RD, Buck EC, et al. (2005) Radiation damage effects in candidate titanates for Pu Pu disposition: Pyrochlore, *J Nucl Mater* 345: 109-135.
25. Begg BD, Hess NJ, Weber WJ, et al. (2001) Heavy-ion irradiation effects on structures and acidic dissolution of pyrochlores. *J Nucl Mater* 288: 208-216.
26. Zhang Y, Hart KP, Bourcier WL, et al. (2001) Kinetics of uranium release from Synroc phases. *J Nucl Mater* 289: 254-262.
27. Ewing RC, Haaker RF, Lutze W (1982) Leachability of zircon as a function of alpha dose, in *Scientific Basis for Nuclear Waste Management V.*, Lutze W (ed), Elsevier, New York, 389-397.
28. ANDRA (2005) Dossier 2005 Argile: Evaluation of the feasibility of a geological repository in an argillaceous formation. December 2005 and Dossier 2005 Granite: Assets of granite formations for deep geological disposal.
29. Lumpkin GR (2001) Alpha-decay damage and aqueous durability of actinide host phases in natural systems. *J Nucl Mater* 289: 136-166.
30. Geisler T, Schaltegger U, Tomaschek F (2007) Re-equilibration of zircon in aqueous fluids and melts. *Elements* 3: 43-50.
31. Tromans J (2006) Solubility of crystalline and metamict zircon: A thermodynamic analysis. *J Nucl Mater* 357: 221-223.

2

Prenatal exposures and exposomics of asthma

Hyunok Choi [1], Mark T. McAuley [2] and David A. Lawrence [3]

[1] Departments of Environmental Health Sciences, Epidemiology, and Biostatistics, University at Albany, School of Public Health, One University Place, Rm 153, Rensselaer, NY 12144-3456, USA
[2] Chemical Engineering Department, Faculty of Science and Engineering, Thornton Science Park, University of Chester, Chester, CH2 4NU, UK
[3] Wadsworth Center, New York State Department of Health, Center for Medical Sciences, Rm 1155, 150 New Scotland Ave. Albany, NY 12201, USA

Abstract: This review examines the causal investigation of preclinical development of childhood asthma using exposomic tools. We examine the current state of knowledge regarding early-life exposure to non-biogenic indoor air pollution and the developmental modulation of the immune system. We examine how metabolomics technologies could aid not only in the biomarker identification of a particular asthma phenotype, but also the mechanisms underlying the immunopathologic process. Within such a framework, we propose alternate components of exposomic investigation of asthma in which, the exposome represents a reiterative investigative process of targeted biomarker identification, validation through computational systems biology and physical sampling of environmental media.

Keywords: air pollution; prenatal; early-life; in utero; exposome; asthma; immune; metabolomics; indoor

---

1. Introduction

Childhood asthma represents a heterogeneous set of conditions, in which, functional lung impairment, chronic inflammation, tissue remodeling, and response to therapy represent some hallmark events. In spite of phenotypic differences, a number of international guidelines point to airway inflammation control as the primary goal of therapy [1]. Another key event of asthma features altered development of the cellular and molecular components of the immune system prior to the appearance of the asthmatic phenotype [2].

1.1. Burden of childhood asthma

Allergic asthma, allergic rhinitis, and eczema represent the most common childhood chronic diseases in the industrialized world [3]. Worldwide, the mean prevalence of asthma is estimated to range between 5 and 25l for 6–7 age group [4], reflecting plateauing of prevalence within many industrialized countries. Worldwide, the mean prevalence of asthma is estimated to range between 5 and 25l for the 6–7 age group [4], reflecting plateauing of prevalence within many industrialized countries [4,5]. Within United States (US), an estimated 13 □0.4% of African American, 13 ± 1.7l of American Indian Alaska Native, and 8 □0.2 l of White children suffer from asthma [6].

The economic toll of living with asthma is estimated at $37.2 billion for 2007 in the US alone[7]. An estimated 10.5 million school days were missed due to asthma between 2005–2009 [8]. Furthermore, the burden of asthma is disproportionately greater for low income and minority children[9].

In spite of intensive on-going investigations on genetic [10], environmental [11], and lifestyle-related [9] risk factors, true causal processes remain unknown [12]. Known multiple classical allergens and adjuvants to date, including infiltrated ambient air pollutants, cockroach and cat dander, dust mites, mice allergens, and environmental tobacco smoke (ETS), do not adequately explain the burden of asthma in children [13,14]. Furthermore, current efforts have failed to stem the growing worldwide development of childhood asthma [15,16]. To date, early-life mechanisms underlying the childhood asthma phenotype remain inadequately defined [15,16,17].

1.2. Aim and scope of this review

Within this review, we examine the early-life exposure to emerging classes of environmental pollutants and their role in respiratory and immune system impairment. As pregnancy is often associated with longer daily hours spent within the indoor environment, the scope of this review is limited to environmental pollutants that pregnant women are exposed to within their residential indoor environment. Within this context, we also examine limited sunlight exposure and the resulting low level of vitamin D as a behavioral sequela of large daily hours spent within the indoor setting. Therefore, we focus here is the combined contributions of gestational vitamin D deficiency and indoor environmental pollutant exposures on pre-clinical immune functional impairments during the first few years of a child's life. Specifically, our goal is to assess the internal validity of exposomic approaches to clarify the mechanisms underlying developmental impairments of the immune system. Here we are concerned with the segment of the population which includes non-occupationally exposed pregnant women with a low risk of adverse birth outcomes. Thus, specific occupational exposures to physical or chemical respiratory risk factors lie beyond the scope of this analysis. Furthermore, classical risk factors, including specific genetic susceptibility contributors, viral, fungal and bacterial agents within the indoor environment, maternal nutritional qualities (such as low dietary intake of vitamin D), social- and or economic- stresses, have already been examined in a number of other excellent reviews[11,18,19,20]. The risks of classical allergens and adjuvants will not be assessed within the present review.

Based on the above considerations and within the context of asthma investigations, we assess the applicability and adaptability of the "exposomic approach" as a set of tools for a comprehensive exposure history characterization and as a health diagnostic tool for parents and infants. As previously defined [21], the exposome is all exogenous factors (the environment) that can modify the endogenous host characteristics (genes and metabolic activities) that together influence health. The endogenous influences have been expanded to include the microbiome and the chimeric associations of pregnancy. For example, children with autism spectrum disorders (ASD) as well as maternal allergies, matenal psychological and socioeconomic difficulties, chemical, and biological (e.g. infection) stress during the pregnancy could contribute to increased risks of airway symptoms [22]. Interestingly, boys (6–8 years of age) have more doctor-diagnosed asthma and ASD than girls [22].

This review is composed of three sections. First, we lay the contextual framework of our discussion in terms of prenatal exposure to air pollution within a home indoor setting and its risks on asthma. Secondly, we examine the state of knowledge regarding exposure and outcome relationship. Thirdly, we propose novel components for exposomic investigation in light of the current gaps in knowledge regarding exposomic approaches as well as the future research direction for asthma biomarker validation.

2. Methods

A PubMed search was conducted using the search terms, asthma, immune, inflammation, oxidative stress, reactive oxygen species, lung, impairment, indoor, air pollution, prenatal, and childhood for peer-reviewed, primary research publications in the English language.

3. Results and Discussion

3.1. Home: context of in utero exposures

substantial body of epidemiological evidence demonstrates that the indoor environment during early-life is critical to the occurrence of childhood asthma [11,23,24]. Children whose parents or caretakers report qualitative indications of indoor dampness, mold, poor ventilation, and recent home redecoration are at significantly greater risks of asthma and allergies [13,24,25]. Exposure within the indoor environment is compounded by energy conservation efforts for buildings, which reduce air exchange rates and promote indoor moisture buildup [26]. Home indoor dampness is estimated to be present in¨ 70l of the homes in upstate New York [27] and in¨ 50l of the homes in Europe [28]. Both adults and children spend an estimated¨ 90% of their daily hours within an indoor setting [29]. Within a vulnerable segment of the population (e.g. elderly, pregnant women, and young children), the exposure to indoor toxicants are estimated to be higher because of sedentary behavior[29]. However, while qualitative reporting of dampness is an established risk factor of asthma, home indoor dampness overall poorly correlate with directly measured fungi, mycotoxins, and other biogenic markers [30]. Furthermore, a growing body of investigations has failed to demonstrate an association between directly measured fungal components and asthma [14,24]. Recent reviews point out the critical need for the identification of specific etiologic agents underlying the surrogate marker of dampness in asthma and allergy investigation [11,25].

The exact mechanism through which a damp environment increases the risk of asthma inception and other lower airway allergic diseases remains unknown [23,24]. Another emerging line of evidence suggests home indoor dampness is correlated with modern chemicals [31]. Within a non-occupational setting, both new and degrading building structural material with polyvinyl chloride (PVC) has shown to initiate or exacerbate asthma [32,33,34]. Sources such as PVC flooring or phthalate concentration in dust are significantly associated with a risk of asthma diagnosis [22,35]. Occupational exposure to painted surfaces have long been associated with eye irritation, skin itching, obstructive airway problems, and frequent urination in longitudinal follow-up studies [36,37,38,39,40,41].

However, exposure assessment of specific chemicals remains challenging. Data regarding temporal variability of the sources, emission, and human behaviors, and resulting human exposure remain scarce. The home indoor volatile organic compound (VOC) concentrations could persist or regenerate within the home indoor environment, aided by poor ventilation, humidity, and or temperature variation in the indoor setting.

3.1.1. Propylene glycol and glycol ethers, the forgotten endocrine disruptors

A robust body of literature suggests that volatile organic compounds associated with cleaning tasks increases the risk of doctor-diagnosed asthma. Among the occupationally exposed adults, indoor sources such as cleaning spray, cleaning liquids, mechanical floor scrubbing, and window cleaning were significantly associated with risks of newly-onset asthma and other respiratory diseases[13,37,42,43,44,45]. Within a non-occupational setting, the children, whose homes had the highest 25l of propylene glycol and glycol ethers (PGEs) concentrations, had a 2.0-fold greater likelihood of having doctor-diagnosed asthma (95l CI, 0.9–4.4), a 4.2-fold greater likelihood of rhinitis (95l CI, 1.7–10.3), and a 2.5-fold greater likelihood of eczema (95l CI, 1.1–5.3) [46].

PGEs are globally distributed as organic solvents and coalescents within cleaning agents, paints, pharmaceuticals, inks and consumer products (e.g., cosmetics, PVC, and glue) [40,47,48]. In various animal models, PGEs cause male sub-fertility and infertility, an increased time to pregnancy [49], oocyte depletion, spontaneous abortion, hematopoietic, immune (e.g., thymic) function suppression in adults, and transplacental fetotoxicities [48].

3.1.2. Other volatile organic compounds of mistaken origin

To date, a great deal of confusion exists regarding the true sources of common VOCs of indoor origin [50,51]. For example, 1-octen-3-ol and 1-butoxy-2-propanol are traditionally known as microbially emitted VOC [50]. However, recent work suggests that they are more likely to be emitted from PVCs at home, glues and other building structural material To date, a great deal of confusion exists regarding the true sources of common VOCs of indoor origin [50,51]. For example, 1-octen-3-ol and 1-butoxy-2-propanol are traditionally known as microbially emitted VOC [50]. However, recent work suggests that they are more likely to be emitted from PVCs at home, glues and other building structural material [36]. Furthermore, 2-ethyl-1-hexanol is traditionally known as MVOC and a "sick building syndrome" compound [52]. However, 2-ethyl-1-hexanol could also be produced as a by-product of di-2-ethylhexyl phthalate (DEHP) degradation [53]. 2-ethyl-1-hexanol from newly painted home indoor surfaces are significantly associated with asthma symptoms and airway inflammatory symptoms in a population-based survey of adults [52]. Another common plasticizer, 2,2,4-trimethyl-1,3-pentanediol diisobutyrate (TMPD-DIB, also known as TXIB) possesses adjuvant properties, including allergic airway inflammation, heightened production of IL-12, and oxidative stress in allergen sensitized mice [32]. 2,4-trimethyl-1,3-pentanediol monoisobutyrate (TMPD-MIB, also known as Texanols™) is globally distributed coalescing agents in latex (i.e. water-based) paints, print-ink, and PVC flooring material [54]. Non-occupational exposure to TXIB and Texanols™ in the general population is a concern because they are detected in considerable concentrations in ambient air in the city of Los Angeles [54], newly painted and installed housing units [55,56,57], and in food packed in polystyrene and polypropylene cups [58]. Furthermore, both TXIB and Texanols™ represent some of the most common VOCs within the temporary housing units constructed by the US Federal Emergency Management Administration for the families displaced by Hurricane Katrina in 2005 [56].

3.1.3. Secondhand smoke exposure(SHS)

SHS is a well-known risk factor for asthma [59]. Second-hand smoke (SHS) is composed of 4000 chemicals, some of which are mutagens carcinogens, developmental toxicants, and irritants [60]. Prenatal exposure to ETS has been associated with adverse effects on fetal and child growth [61] and cognitive functioning during childhood [62], as well as exacerbation of childhood asthma and with allergic response [63]. Furthermore, SHS is a primary delivery mode of other VOCs, including benzene, toluene and styrene in the US population [64].

3.1.4. Vitamin D

Higher confinement to indoor life (and associated deprivation from sunlight) can lead to vitamin D insufficiency with concomitant increases in exposure to indoor pollutants. A vitamin D deficiency is associated with chronic lung disease [65], which likely relates to less control of the inflammation and oxidative stress induced by the indoor air. Vitamin D insufficiency (serum level of 25(OH)D 30 ng mL) is common among children in the United States [66,67]. Among children aged 6–11 years, vitamin D deficiency was approximately 62l in non-Hispanic whites, approximately 86l in Hispanics, and approximately 96l in non-Hispanic blacks [66]. Prior work has shown that Vitamin D deficiency leads to impaired Th1 and Th17 lymphocyte function [68]. Furthermore, it has also been associated with reduced activity of FoxP3 functions in T_{reg} cells [68]. Low serum 25(OH)D has been strongly implicated in asthma due to its correlation with oxidative stress, resultant tissue damage, and airway inflammation [69]. In addition, cytokine production of cord blood T cells is correlated with season [67,70]. Thus, season could influence the exposure to home indoor toxicants by promoting the family members to reduce the home ventilation rate, nutritional intake of food items rich in vitamins (e.g., spinach, kale, collards, soy beans, and rainbow trout), sunlight (thereby, vitamin D) or maternal infection, as well as the asthma outcome [67,70].

3.2. Childhood immune markers and asthma risks

3.2.1. A definition of immune system impairment

Developmental alterations during early life in the immune system is thought to underlie subsequent allergic diseases [71]. A hallmark of such pre-clinical events consists of preferential enhanced development of the CD4 type-2 helper T cells (Th2) [72]. Within this review, we narrow the scope of altered immune system development as skewed T-cell differentiation, including Th1, Th2, Th17 and regulatory T (T_{reg}) cells, and innate immune cells.

3.2.2. Biomarkers of immune system impairments

Several indoor environmental exposures to modern chemicals (from building material, consumer product uses, and lifestyle) within a damp home indoor environment (contributed by structural water damage, poor ventilation, and or low temperature) have been shown to contribute to altered development of fetal innate and adaptive immune cells [34,73,74,75,76]. A critical event in this process is the unbalanced development of Th1 and Th2 cells [72]. The Th2 phenotype naturally dominates during fetal development, but environmental pollutants are suggested to maintain or enhance this imbalance after birth. Accordingly, unclear events preceding Th1 Th2 imbalance and identification of high risk newborns represent key barriers to our understanding of this process [71]. An improved understanding of prenatal exposure to indoor synthetic pollutants and the risks on prolonging the Th1 Th2 cell imbalance represents the main challenge to aid interventions, which limit risks during pregnancy.

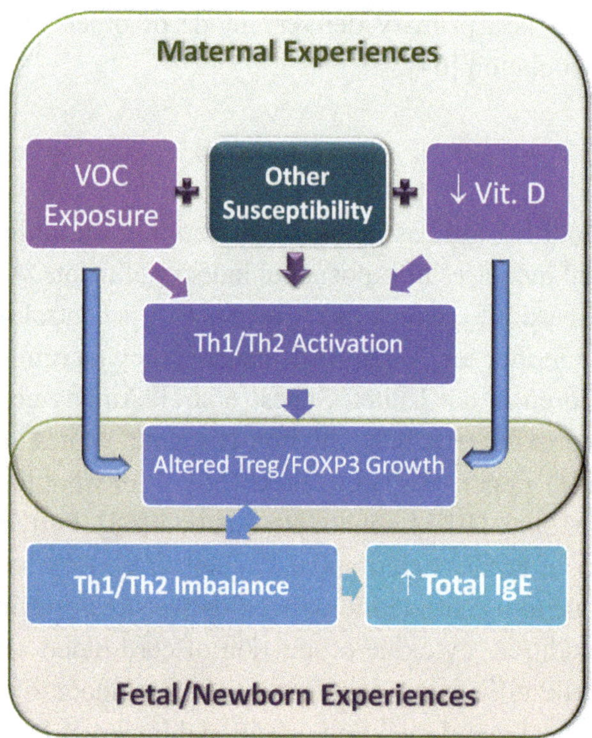

Figure 1. Proposed conceptual model.

A conceptual framework of our work is outlined in Figure 1. Multiple-hits on the maternal immune system from home indoor exposure to VOCs, other susceptibilities (i.e., maternal anthropomorphic traits, maternal and paternal atopy, secondhand smoke, and season), and low vitamin D level, activate innate immune cells (e.g., NK cells, macrophages, neutrophils). Such multiple-hits promulgate inflammation and resultant oxidative stress, further skewing maternal T cells toward Th2 development [77], while suppressing T regulatory (T_{reg}) cells and the associated Foxp3 gene expression [78]. Suppression of maternal T_{reg} cells, in turn, leads to an unbalanced development of cytokines produced by the Th1, Th2 cells as markers of pre-clinical, immunotoxic consequences in newborns. Here, we posit that Th2 cells, which release interleukin (IL)-4, IL-5, and IL-13 promote asthma. Increased IL-4 production heightens the IgE level, which arms mast cells and basophils for their detrimental allergen-induced release of bronchoconstrictor mediators [79]. IL-5 increases the presence of eosinophils [79] and IL-13 increases bronchoconstriction. Therefore, environmental factors that cause a Th1-Th2 imbalance in the prenatal or early postnatal period, in particular, T-cell polarization toward Th2 reactivity, may contribute to the risk of allergic disease. Genetic associations have also connected asthma, inflammation, and atopy [80], which further supports our proposed analysis of the associations of environmental exposures, immune biomarkers, and asthma severity.

3.3. Exposome

There are three biosystems (maternal, fetal, and placental), which need to be considered for the exposomic investigation of prenatal effects. Each of these biosystems has an influence on the health of offspring for their lifetime after maternal exposure to environmental stressors. The placenta is grouped as a separate entity, because it is not a permanent organ of either the mother or offspring, it is an intertwined mix of maternal and fetal cells, and its physiological presence is vital for fetal existence. The lifetime consequences of developmental exposures modulate both the immunity [81,82] and behavior of offspring [83,84]. The exposomal composite [21,85,86] begins at conception, but the constant and consistent interplays of environment and genetics affect health throughout the lifetime. Developmental exposomics includes the analyses of all exogenous environmental (biological, chemical, and psychological physical) stressors on these three biosystems and the resultant effects on endogenous cellular and molecular activities, which affect health. These include the body's dynamic adaptation, repair, or further damage in response to the varying environmental influences [87]. The effects of prenatal exogenous stressors and the endogenous responses to the environmental modulators, which depend on the fetomaternal genotypes, determine whether a disorder disease will occur with greater prevalence within the lifetime of the offspring. The exposome paradigm suggests that pollutants must be evaluated with consideration of the additive, synergistic and "potentiated" affects from all environmental influences, including diet and life-style behaviors, which includes psychological and physical stress, and that the environmental influences are dependent on each person's genetics [88]. Consideration of potentiated effects are especially important for exposomic analysis since two different stressors may induce no detectable effect on separate pathways, but these pathways converge leading to a detrimental response [84].

As the fetus develops, the metabolomics [89] and proteomics of the mother, placenta (the interconnecting fetomaternal barrier) and fetus regulate the constituents for fetal growth and differentiation. At conception and later in gestational development, environmental stressors, including air pollutants, have been reported to affect the prevalence of asthma in the offspring [90,91]. It seems somewhat incongruous that an air pollutant, which affects the maternal lung could affect fetal development. However, some pollutants may directly permeate to the placenta and fetus and other pollutants may induce endogenous maternal cellular and molecular changes that transmit the lung exposure to the placenta and fetus [92,93,94]. It is especially interesting to note that mast cells in the lung interstitium and alveolar epithelium and in close proximity to the vascular and nervous system networks are implicated in asthma, because mast cells also play an important role in placenta development. The airways of asthmatic individuals have increased presence of mast cells, and mast cells have been suggested to enhance the asthmatic pathology [95,96,97]. Mast cell release of histamine and proteases occurs with ozone exacerbation of asthma [98,99,100,101,102]. Mast cells are often considered to play a role in allergic asthma with regard to antigen-specific allergen triggering via IgE. However, mast cells and other innate immune cells can enhance asthma with pattern recognition receptor (PRR)-induced release of prostaglandins and other inflammatory factors due to exposure to pathogen-associated molecular patterns (PAMPs). The involvement of innate immune cells in asthma may be one reason why inhaled corticosteroids (ICS) sometimes have little effect, in that ICS have minimal influence on the number of mast cells in asthmatic lungs [103] and on bronchoconstriction[104].

Air pollutants can modulate placenta structure and function [105,106,107,108,109], including immune activities associated with the placenta [110]; thus, it is not surprising [110]; that they either directly affect the placenta and fetus or rely on transited maternal cells and molecules. Hematopoietic mast cells have been suggested to play a role in implantation and placenta development [111,112], which affects offspring birth weight. Mast cell activity and other immune activities in the placenta affect placenta size, which can affect birth weight. Mast cells [113] and air pollutants [113] and air pollutants [114,115,116] also have been linked to preeclampsia. The placenta controls bidirectional trafficking of cells and molecules, and thus plays an intermediary role in the fetomaternal relationship.

Asthma is a multifaceted inflammatory disease of the airways that usually includes increased mucus production, bronchorestriction, and airway remodeling. Whether asthma is classified as allergic or non-allergic immune cells producing similar cytokines and chemokines are implicated. Allergic asthma involves mainly adaptive lymphocytes including CD4 Th2 cells producing IL-4, IL-5, and IL-13. IL-4 promotes B lymphocyte switching to the IgE antibody isotype, which arms mast cells and basophils with antigen-specific signaling. IL-5 promotes the growth and differentiation of eosinophils, and IL-13 promotes contraction of airway epithelial cells and smooth muscle cells. Non-allergic asthma involves innate lymphocytes of different classes referred to as innate lymphoid cells (ILC)1, ILC2 and ILC3 [117], which respond to PAMPs with their PPRs and produce IL-4, IL-5, IL-13 and IL-17. The fetomaternal condition skews both the mother's and fetus's helper T cells toward the Th2 phenotype via epigenetic modulations as the na□ve T cells are induced to proliferate. As the offspring gets exposed to the non-aseptic environment at birth further epigenetic modulation generally leads to a more balanced ratio of Th1 and Th2 cells; however, prenatal and postnatal environmental influences, including ozone, cigarette smoke, diesel exhaust and other airborne pollutants seem to enhance the continuation of the Th2 dominance [118]. With regard to allergic asthma the enhanced maintenance of the Th2 imbalance may be due to the phenotype of the antigen presenting cells (APCs). Alternatively, activated macrophages or M2 macrophages or dendritic cells preferentially activate Th2 cells. Oxidative stress from pollutants can lower the amount of cellular glutathione (GSH, the cell's main anti-oxidant) and APCs with less GSH preferentially activate Th2 cells [118,119,120]. Interestingly, GSH loss has been implicated in the induction of asthma due to numerous airborne metabolites exposures that can generate oxidative stress [121]. Prenatal exposure to acetaminophen also influences the risk of wheezing incidence at age 5 due to GSH variance [122]. Oxidative processes from prenatal exposures to cigarette smoke [123,124], polycyclic aromatic hydrocarbons [125], PCBs [126], phthalates [127], and bisphenol A [128] are likely due to aryl hydrocarbon receptor (AHR) activation[129], which promotes inflammation and asthma; these pollutants are known to activate AHR [129,130].

Allergic asthma is often of low or moderate severity with a relatively low level of airway remodeling although it is as chronic as the more severe form. Severe asthma with more recurrent exacerbations usually involves a greater extent of inflammation with a greater presence of neutrophils and eosinophils. Severe asthma also includes more IFNγ and IL-17 from Th1 and Th17 cells or ILC2 and ILC3 cells. The ILC2 and ILC3 subsets are lineage negative cells; whereas, the ILC1 subset includes NK cells. ILC2 and ILC3 cells are IL-7R positive; IL-7 is produced by hematopoetic stromal cells, and it supports early lymphoid progenitor differentiation of the stem cells. The ILC2 population is very similar to the Th2 population with regard to its cytokine profile; it develops in the presence of IL-25 and IL-33, and like Th2 cells, it expresses the GATA-3 transcription factor. IL-33 is a nuclear protein released by airway epithelial cells damaged by airborne pollutants; it is referred to as a damage-associated molecular pattern molecule (DAMP) or as an alarmin and is activated by cleavage with mast cell protease [131]. Th2 and ILC2 cells, mast cells, basophils and eosinophils are stimulated by IL-33. Another stimulator of ILC2 cells is thymic stromal lymphopoietin (TSLP) [132], which like IL-33 is released from airway epithelial cells and smooth muscle cells [133]. Damaged airway cells also release ATP [133]. Damaged airway cells also release ATP, which recruits and activates mast cells[134,135]; via ATP-induced activation of the inflammasome, cleaved active form of IL-1β is produced [136], which also recruits neutrophils and Th17 cells, and inflammation and airway remodeling is enhanced [137].

3.3.1. Utility

Attraction and promise of Exposome-Wide Association Studies (EWAS) lie in distinguishing biomarkers of exposure from biomarkers of disease phenotype [88,138]. Recent blood-based exposomic analyses have shown their potential to capture multiple environmental factors, including the nutrients [139], viral infection [140], the microbiome [141], stress [142], and environmental toxins[143]. For example, preterm delivery as well as fetal growth restriction has been shown to be associated with urinary levels of acetate, tyrosine, formate, trimethylamine, lysine and glycoprotein [144].

Rappapport (2012) proposes a two-phase strategy for exposomic investigation of the disease outcomes. Within the first phase, molecules are scanned through an agnostic search tool for their association with the disease outcome status [138]. The second stage investigation is often directed to investigate the role of the metabolites (identified in the first phase) as causal contributors or intermediate players in pathophysiology of the clinical phenotypes [145]. The first phase of the analyses is often conducted through liquid-chromatography tandem mass spectrometry (LC-MṢ MS) or nuclear magnetic resonance (NMR) spectroscopy [146]. Each has advantages and shortcomings. LC-MS offers the advantage of greater sensitivity of molecules in biofluids (e.g.,′ 4000 molecules within plasma sample) and better molecule discrimination [146]. On the other hand, NMR detect[146]. On the other hand, NMR detects relatively smaller set of particles (200). But, the NMR does not destruct the sample, thus permitting repeated analysis [146]. Furthermore, application of LC-MṢ MS is also limited by the fact that existing libraries are too small to identify most of the detected molecules[147]. To date, interpretability of many metabolomics analyses remain extremely limited due to fragmentations or adduct formations [148]. To date, statistical methodology regarding pattern recognition remains the major challenge associated with metabolomics analysis. The challenge lies not only in the association of biomarkers with the disease outcome of interest, but also in clarifying the relevance of a given biomarker in a pathophysiological process [145]. Given the preliminary nature of small molecular biomarkers for phenotypic analyses, independent validations in order to account for the high-dimensionality of the analyses and the associated false discovery rates represent critical steps in identification of true disease outcome markers [138].

3.3.2. Urinary metabolomics

Metabolomics captures a comprehensive catalogue of small molecules (molecular weight.. 1500 Daltons) within any organic systems or physiological state [149]. Urine metabolomic profiling holds promise as a non-invasive and unbiased window to the "sum of history" of the present disease status for newborns and children [145]. For example, within a murine model of asthma, administration of dexamethasone induced a marked, yet, reversible metabolism of carbohydrate, lipid, and sterol in bronchoalveolar lavage fluid from asthmatic lungs [149].

Within a clinical setting, the urinary metabolic profile of preterm and term infants were distinctly unique according to the gestational age at delivery [150]. In particular, preterm status was associated with altered patterns of: 1) tyrosine metabolism; 2) the biosynthesis process of tyrosine, tryptophan, and phenylalanine; 3) urea cycle; 4) arginine metabolism; and 5) proline metabolism [150]. Furthermore, co-morbid condition of extremely low weight as well as preterm delivery status was associated with altered arginine-proline metabolism, purine-pyrimidine metabolism, and urea cycle during adulthood[150]. In another pilot investigation of children with nephron-uropathies (renal dysplasia, vesico-ureteral reflux, urinary tract infection, or acute kidney injury) versus healthy children, the disease status was associated with an alteration in the urea cycle as well as purine and pyridine synthesis [150]. Metabolic profiling of asthmatic versus healthy control children using liquid chromatography—mass spectrometry (LC-MS) demonstrated lower levels of urocanic acid and methyl-imidazoleacetic acid in the asthmatic children compared to the controls [148]. Both urocanic acid and methyl-imidazoleacetic acid play a role in inflammatory responses. Overall, the number of studies for the identification of the biomarker using this technology remains extremely limited due to the limitations in analytical chemical techniques as well as the computational and bioinformatic tools[145].

3.3.3. Predictive value of metabolomic profile for asthma

Undirected metabolomics refers to the comprehensive measurement of all small molecules within a given sample. Such data is subsequently cleaned into a small set of signals and qualitatively recognized through *in silico* searches of existing libraries or ionization experiments [151]. While this approach has [151]. While this approach has potential for the identification of novel biomarkers, a number of potential issues with such biomarkers have also been identified [145]. One of the critical issues relate to the erroneous identification of noise rather than a signal [145]. Even when a meaningful signal is detected, the biological relevance of the signal to underlying etiological process remains a challenge [145]. For example, Nuclear Magnetic Resonance (NMR) based metabolomic analysis applied to exhaled breath condensate demonstrated greater sensitivity to discriminate asthmatic children (86l sensitivity) compared to that based on exhaled nitric oxide and Forced expiratory volume (FEV_1) with 81l sensitivity [152]. However, the exact molecular identity could not be determined based on the NMR technique [152]. Thus, the issues of biological plausibility, reproducibility of the identified set (through second stage analysis) as well as confounding need to be addressed [145]. To date, most investigations have been able to identify only the biomarkers of clinical outcome status (e.g. cardiac ischemia) [145]. Thus, the development of predictive metabolomics biomarkers of disease outcome is needed as diagnostic tool.

In contrast, directed metabolomics refers to a more focused search of selected set of known and expected metabolites using tandem mass spectrometry (MS MS), or selected ion recording (SIR) with GC-MS [151]. Some investigators have used the targeted search as a validation tool for preliminarily identified set of metabolites [145].

3.3.4. Computational systems biology and its application to asthma

Computational modelling includes an array of quantitative techniques that can be used to assess health risk. The computational modelling offers the advantage of interpreting preliminarily identified signal in the second stage of exposomic investigation. Furthermore, the computational models could account for the underlying uncertainties inherent within complex biological systems and their response to environmental exposure. The type of model depends on the nature of the system to be modelled[153]. Historically, the physiologically based pharmacokinetic [153]. Historically, the physiologically based pharmacokinetic pharmacodynamic (PBPK PD) model has been routinely employed in toxicological studies [154]. This approach uses an organism's toxicokinetic information to evaluate normal tissue limitations to chemical exposure (Figure 2). These models need information on the pollutants, including its absorption, distribution, metabolism, and elimination (ADME). In addition, information about the organ size, ventilation rates, the age, and the gender of the organism,

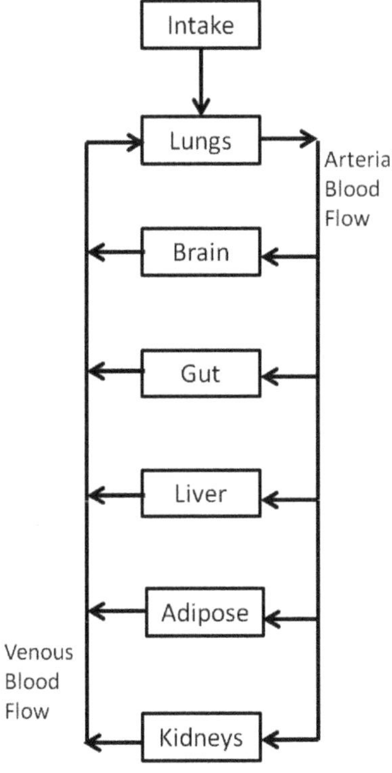

Figure 2. Generic structure of the PBPK modelling framework.

together with the hypothesized mechanistic information are required. PBPK modelling in early life has been used to determine toxicant exposure in children [155,156], however the focus has centered on drug metabolism and there is a paucity of models, which focus on environmental exposure and prenatal risk of developing asthma. In terms of prenatal exposures to toxicants, PBPK models have been used to quantify and characterize the physiological effects of drug exposure during pregnancy [157]. A noteworthy development in PBPK modelling is the development human PBPK model toolkit by recoding published PBPK models [158]. This toolkit is expected to provide a straightforward reference point to a wide variety of models of this nature. The last decade and a half has also witnessed the growth of the systems biology paradigm as a framework for conducting toxicological investigations[159]. At the core of this new approach is computational systems biology [160]. The aim of computational systems biology is to use mathematics to describe complex biological phenomena in a quantitative manner using computer simulations (Figure 3). There is also an emphasis on using the models to study biological systems in a holistic manner as opposed the more reductionist approach that is often adopted in the wet-laboratory [161]. Several different mathematical approaches can be used when designing a computational systems model. These different mathematical approaches were recently reviewed in detail [162]. By far the most ubiquitously adopted mathematical approach is to use ordinary differential equations (ODEs). This method is similar in nature to PBPK models, in that, model reactions are informed by kinetic data and steady-state information about the biological system

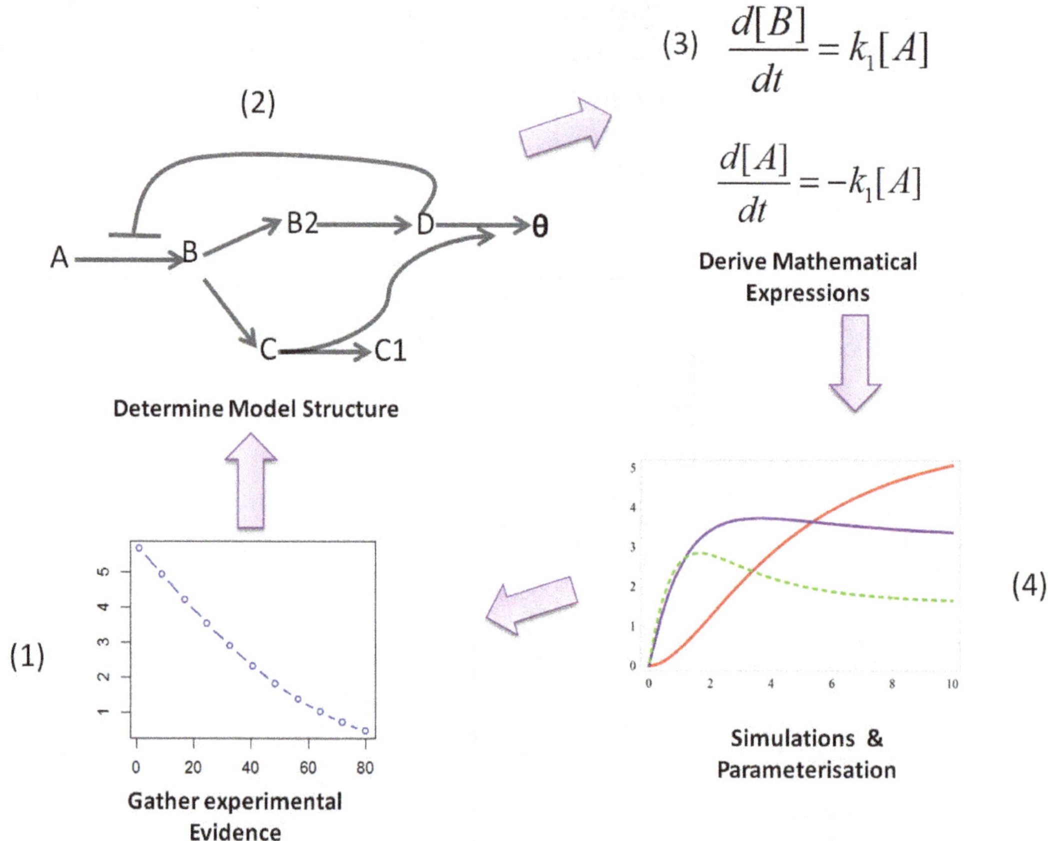

Figure 3. The steps involved in constructing a computational systems biology model.

under investigation. A disadvantage of these models is that they cannot include variability or spatial dynamics as part of the behavior of the biological system. Another disadvantage is that kinetic information is often difficult to obtain, despite a number of archives for online kinetic data [163]. Parameterizing such models is also challenging especially as the size of model increases [164]. A pertinent example of how an ODE model can dovetail with wet-laboratory experimentation is recent work by Carbo et al. (2014) which was used to investigate the role of interleukin-21 (IL-21) in the gastric mucosa during *H. pylori* infection [165]. Computational systems modelling has been used to model CD4 T cell differentiation with *in vivo* mechanistic studies. The model predicted activated expression of T-bet and RORγt and the phosphorylation of STAT3 and STAT1 and suggested a potential role of IL-21 in the modulation of IL-10. This combined approach indicated that IL-21 regulates Th1 and Th17 effector responses during chronic *H. pylori* infection in a STAT1- and STAT3-dependent manner, and are thus regulators of the *H. pylori* infection. Of significance for this review is that it suggests *H. pylori* colonization protects against childhood asthma [166,167]. Therefore, this study outlines a way in which the immune system can be studied mechanistically by using a combined *in vivo in silico* approach. Stochasticity in systems models can be dealt with by using stochastic differential equations which represent biological reactions as discrete random molecular collisions. Such models often represent scenarios involving low molecule numbers such as protein molecules from an expression of a single gene [168]. The major disadvantage of such models is that they are computationally slow. Variability can also be represented by using Bayesian networks. This type of modelling is useful for making toxicant risk estimates [169]. Petri nets can also be used to construct a systems model. These are a special type of model which comprises of two nodes, called places and transitions. Places and transitions are connected using arrows. Each place contains a number of tokens which is the same as a discrete number of biochemical molecules. A Petri net functions by input- output firing at the 'transitions' within the network. The 'firing' of transitions represents a reaction taking place within a biological system [170]. There are a number of different standards for exchanging computational models, with the Systems Biology Markup Language (SBML) leading exchange format for the exchange of biological models at present [171]. This format is designed to be used for exchange independent of the modelling tool used to construct the model. Another more recent development from a purely toxicological perspective is the development of adverse outcomes (AOPs) [172]. AOPs are in essence are a way of having a mathematical representation which is informed by our current understanding of a biological pathway and how a change to that pathway could have an adverse outcome at the level of a whole organism or even population level [172]. Of note for this review is that this novel modelling methodology has been applied to studies which have investigated respiratory allergies [173]. The stages in building an AOP include determining the information required to represent the pathway of interest. A summary is then included in the form of a flow chart that depicts the AOP from molecular initiation even to its effect on the whole organism. The weight of the interaction is determined together with a confidence determination [174]. In light of developments such as those outlined in this section computational modelling will be applied further to both pre and post natal exposure to chemicals investigations.

3.3.5. Is external validation necessary within exposome

In contrast to biomarker driven definition of exposome [138], the US National Research Council defines exposome as "the extension of exposure science from the point of contact between stressor and receptor inward into the organism and outward to the general environment, including the ecosphere"[175].

Within the context of asthma investigation, the more holistic definition proposed by the National Research Council is needed for the following reasons.

First, since most metabolomics studies have been of pilot-scale, validation of the biomarkers both within- as well as between-persons remains exceedingly rare. Furthermore, underlying representativeness of the diseased versus healthy controls remains relatively poorly understood. For example, age-dependent appearance of the asthma phenotype requires validation of initial identification.

Second, temporal variability in external and internal exposures remains unknown [176]. At best, they are measured with substantial error [176]. Lack of valid and accurate exposure assessment data represents an important barrier to the understanding of true human exposure [11,13]. The nature of specific pollutants and the concentrations of presumed personal exposures remain unknown. In spite of these, predominant methodological approaches of asthma investigations rely on cross-sectional design[35,177,178,179]. Cross-sectional design is inherently limited in its ability to reject the possibility that risk factors (e.g. water-based cleaning) were adopted following the health outcome occurrence. In addition, most investigations rely on surrogate exposure markers and memory recall of potential sources (e.g., damp spots in ceiling, specific product uses, the history of renovation, or flooding) rather than direct measurement [13,24,25].

4. Conclusion

An improved understanding of the mechanistic processes underlying immune system impairments, early detection, and source removal are critical for prevention of asthma as a globally burdensome group of illnesses. This review considered damp home environments, emission and or retention of VOCs from cleaning chemicals, latex paints, adhesives, rotting plasticizers, other aging furniture and structural material in homes as a context for exposure during early-life. We propose consideration of context of exposure within the application of targeted exposomic investigations. Such context holds promise for biomarker identification for asthma diagnosis, severity prediction, as well as clarification of pathophysiological processes. We propose an exposome as an integrated investigative approach in which comprehensive environmental media sampling, metabolomic profiling of biological samples, and second-stage investigation of biomarker relevance are considered together in an iterative process.

Conflict of interest

All authors declare no conflicts of interest in this paper.

References

1. Adamko DJ, Sykes BD, Rowe BH (2012) The metabolomics of asthma: Novel diagnostic potential. *Chest* 141: 1295-1302.
2. Wright RJ (2008) Stress and childhood asthma risk: overlapping evidence from animal studies and epidemiologic research. *Allergy Asthma Clin Immun* 4: 29.
3. Committee on the Assessment of Asthma and Indoor Air (2000) Clearing the Air:Asthma and Indoor Air Exposures; Division of Health Promotion DP, Institute of Medicine, editor: The National Academies Press.
4. Asher MI, Montefort S, Bj□rkst n B, et al. (2006) Worldwide time trends in the prevalence of symptoms of asthma, allergic rhinoconjunctivitis, and eczema in childhood: ISAAC Phases One and Three repeat multicountry cross-sectional surveys. *Lancet* 368: 733-743.

5. Asher M, Keil U, Anderson H, et al. (1995) International Study of Asthma and Allergies in Childhood (ISAAC): rationale and methods. *Eur Respir J* 8: 483-491.
6. Brim SN, Rudd RA, Funk RH, et al. (2008) Asthma Prevalence Among US Children in Underrepresented Minority Populations: American Indiaṇ Alaska Native, Chinese, Filipino, and Asian Indian. *Pediatrics* 122: e217-e222.
7. Kamble S, Bharmal M (2009) Incremental direct expenditure of treating asthma in the United States. *J Asthma* 46: 73-80.
8. Akinbami OJ, Moorman JE, Liu X (2011) Asthma prevalence, health care use, and mortality: United States, 2005-2009: US Department of Health and Human Services, Centers for Disease Control and Prevention, National Center for Health Statistics.
9. Bryant-Stephens T (2009) Asthma disparities in urban environments. *J Allergy Clinl Immunol* 123: 1199-1206; quiz 1207-1198.
10. Torgerson DG, Ampleford EJ, Chiu GY, et al. (2011) Meta-analysis of genome-wide association studies of asthma in ethnically diverse North American populations. *Nat Genet* 43: 887-892.
11. Heinrich J (2011) Influence of indoor factors in dwellings on the development of childhood asthma. *Int J Hyg Envir Heal* 214: 1-25.
12. Busse PJ, Wang JJ, Halm EA (2005) Allergen sensitization evaluation and allergen avoidance education in an inner-city adult cohort with persistent asthma. *J Allergy Clinl Immunol* 116: 146-152.
13. Jaakkola JJ, Jaakkola MS (2006) Professional cleaning and asthma. *Curr Opin Allergy Clin Immunol* 6: 85-90.
14. Jaakkola MS, Jaakkola JJK (2004) Indoor Molds and Asthma in Adults. *Advances in Applied Microbiology*: Academic Press. pp. 309-338.
15. Masoli M, Fabian D, Holt S, et al. (2004) Review article The global burden of asthma: executive summary of the GINA Dissemination Committee Report. *Allergy* 59: 469-478.
16. Yang IV, Schwartz DA (2012) Epigenetic mechanisms and the development of asthma. *J Allergy Clin Immun* 130: 1243-1255.
17. Borish L, Culp JA (2008) Asthma: a syndrome composed of heterogeneous diseases. *Ann Allergy Asthma Immunology Today* 101: 1-8.
18. Ho S-M (2010) Environmental epigenetics of asthma: an update. *J Allergy Clinl Immunol* 126: 453-465.
19. Ege MJ, Mayer M, Normand A-C, et al. (2011) Exposure to environmental microorganisms and childhood asthma. *New Engl J Med* 364: 701-709.
20. Hansel TT, Johnston SL, Openshaw PJ (2013) Microbes and mucosal immune responses in asthma. *Lancet* 381: 861-873.
21. Wild CP (2005) Complementing the genome with an "exposome": the outstanding challenge of environmental exposure measurement in molecular epidemiology. *Cancer Epidem Biomar* 14: 1847-1850.
22. Larsson M, Weiss B, Janson S, et al. (2009) Associations between indoor environmental factors and parental-reported autistic spectrum disorders in children 6-8 years of age. *Neurotoxicology* 30: 822-831.
23. Bornehag CG, Sundell J, Sigsgaard T (2004) Dampness in buildings and health (DBH). Report from an on-going epidemiological investigation on the association between indoor environmental factors and health effects among children in Sweden. *Indoor Air* 14: 59-66.

24. Mendell MJ, Mirer AG, Cheung K, et al. (2011) Respiratory and Allergic Health Effects of Dampness, Mold, and Dampness-Related Agents: A Review of the Epidemiologic Evidence. *Environ Health Perspect* 119.
25. Mendell MJ (2007) Indoor residential chemical emissions as risk factors for respiratory and allergic effects in children: a review. *Indoor Air* 17: 259-277.
26. Fang L, Clausen G, Fanger PO (1998) Impact of temperature and humidity on the perception of indoor air quality. *Indoor air* 8: 80-90.
27. Rosenbaum PF, Crawford JA, Anagnost SE, et al. (2009) Indoor airborne fungi and wheeze in the first year of life among a cohort of infants at risk for asthma. *J Expos Sci Environ Epidemiol* 20: 503-515.
28. Eurostat – The Statistical Office of the European Union (2010) Europe in Figures – EUROSTAT yearbook 2010. pp. Chapter 6. Living Conditions and Welfare.
29. Samet JM, Spengler JD (2003) Indoor environments and health: moving into the 21st century. *Am J Public Health* 93: 1489-1493.
30. Institute of Medicine (2004) Damp Indoor Spaces and Health. New York: The National Academies.
31. Weschler CJ (2009) Changes in indoor pollutants since the 1950s. *Atmos Environ* 43: 153-169.
32. B□hisch U, B□hme A, Kohajda T, et al. (2012) Volatile Organic Compounds Enhance Allergic Airway Inflammation in an Experimental Mouse Model. *PLoS ONE* 7: e39817.
33. Bornehag CG, Sundell J, Weschler CJ, et al. (2004) The association between asthma and allergic symptoms in children and phthalates in house dust: a nested case-control study. *Environ health perspect* 112: 1393-1397.
34. Herberth G, Herzog T, Hinz D, et al. (2012) Renovation activities during pregnancy induce a Th2 shift in fetal but not in maternal immune system. *Int J Hyg Environ Health.*
35. Bornehag CG, Sundell J, Hagerhed-Engman L, et al. (2005) "Dampness" at home and its association with airway, nose and skin symptoms among 10 851 preschool children in Sweden: a cross sectional study. *Indoor Air* 15: 48-55.
36. Kim JL, Elfman L, Mi Y, et al. (2007) Indoor molds, bacteria, microbial volatile organic compounds and plasticizers in schools--associations with asthma and respiratory symptoms in pupils. *Indoor Air* 17: 153-163.
37. Wieslander G, Norback D (2010) Ocular symptoms, tear film stability, nasal patency, and biomarkers in nasal lavage in indoor painters in relation to emissions from water-based paint. *Int Arch Occup Environ Health* 83: 733-741.
38. Wieslander G, Norback D (2010) A field study on clinical signs and symptoms in cleaners at floor polish removal and application in a Swedish hospital. *Int Arch Occup Environ Health* 83: 585-591.
39. Wieslander G, Norback D, Edling C (1994) Occupational exposure to water based paint and symptoms from the skin and eyes. *Occup Environ Med* 51: 181-186.
40. Wieslander G, Norback D, Edling C (1997) Airway symptoms among house painters in relation to exposure to volatile organic compounds (VOCS)--a longitudinal study. *Ann Occup Hyg* 41: 155-166.
41. Wieslander G, Norback D, Nordstrom K, et al. (1999) Nasal and ocular symptoms, tear film stability and biomarkers in nasal lavage, in relation to building-dampness and building design in hospitals. *Int Arch Occup Environ Health* 72: 451-461.

42. Wieslander G, Kumlin A, Norback D (2010) Dampness and 2-ethyl-1-hexanol in floor construction of rehabilitation center: Health effects in staff. *Arch Environ Occup Health* 65: 3-11.
43. Wieslander G, Lindgren T, Norback D, et al. (2000) Changes in the ocular and nasal signs and symptoms of aircrews in relation to the ban on smoking on intercontinental flights. *Scand J Work Environ Health* 26: 514-522.
44. Wolkoff P, Schneider T, Kildes□J, et al. (1998) Risk in cleaning: chemical and physical exposure. *Scie Total Environt* 215: 135-156.
45. Zock JP, Plana E, Jarvis D, et al. (2007) The use of household cleaning sprays and adult asthma: an international longitudinal study. *Am J Respir Crit Care Med* 176: 735-741.
46. Choi H, Schmidbauer N, Sundell J, et al. (2010) Common household chemicals and the allergy risks in pre-school age children. *PLoS One* 5: e13423.
47. Ernstg rd L, Lof A, Wieslander G, et al. (2007) Acute effects of some volatile organic compounds emitted from water-based paints. *J Occup Environ Med* 49: 880-889.
48. IARC (2006) Formaldehyde, 2-Butoxyethanol and 1-tert-Butoxypropan-2-ol: Summary of Data Reported and Evaluation. In: ORGANIZATION WH, editor. IARC Monographs on the Evaluation of Carcinogenic Risks to Humans. Lyon: WORLD HEALTH ORGANIZATION.
49. Garlanṭ zec R, Warembourg C, Monfort C, et al. (2013) Urinary glycol ether metabolites in women and time to pregnancy: the PELAGIE cohort. *Environ Health Perspect* 121: 1167–1173.
50. Korpi A, J̈ rnberg J, Pasanen A-L (2009) Microbial volatile organic compounds. *Crit rev toxicol* 39: 139-193.
51. Korpi A, Pasanen A-L (1998) Volatile Compounds Originating for Mixed Microbial Cultures on Building Materials under various. *Appl Environ Microb* 64: 2914.
52. Norback D, Wieslander G, Nordstr K, et al. (2000) Asthma symptoms in relation to measured building dampness in upper concrete floor construction, and 2-ethyl-1-hexanol in indoor air. *Int J Tuberc Lung D* 4: 1016-1025.
53. Nalli S, Horn OJ, Grochowalski AR, et al. (2006) Origin of 2-ethylhexanol as a VOC. *Environ Pollut* 140: 181-185.
54. Goliff WS, Fitz DR, Cocker K, et al. (2012) Ambient measurements of 2,2,4-trimethyl, 1,3-pentanediol monoisobutyrate in Southern California. *J Air Waste Manage* 62: 680-685.
55. J̈ rnstr□m H, Saarela K, Kalliokoski P, et al. (2008) Comparison of VOC and ammonia emissions from individual PVC materials, adhesives and from complete structures. *Environ Int* 34: 420-427.
56. Maddalena R, Russell M, Sullivan DP, et al. (2009) Formaldehyde and Other Volatile Organic Chemical Emissions in Four FEMA Temporary Housing Units. *Environmen Sci Technol* 43: 5626-5632.
57. Wieslander G, Norback D, Bjornsson E, et al. (1997) Asthma and the indoor environment: the significance of emission of formaldehyde and volatile organic compounds from newly painted indoor surfaces. *Int Arch Occup Environ Health* 69: 115-124.
58. Kempf M, Ramm S, Feuerbach T, et al. (2009) Occurrence of 2,2,4-trimethyl–1,3-pentanediol monoisobutyrate (Texanol□) in foods packed in polystyrene and polypropylene cups. *Food Addit Contam A* 26: 563-567.
59. Sly PD, Boner AL, Bj□rksten B, et al. (2008) Early identification of atopy in the prediction of persistent asthma in children. *Lancet* 372: 1100-1106.
60. Couraud S, Zalcman G, Milleron B, et al. (2012) Lung cancer in never smokers – A review. *Eur J Cancer* 48: 1299-1311.

61. Wang L, Pinkerton KE (2007) Air pollutant effects on fetal and early postnatal development. *Birth Defects Res Part C: Embryo Today: Rev* 81: 144-154.
62. Wigle DT, Arbuckle TE, Walker M, et al. (2007) Environmental Hazards: Evidence for Effects on Child Health. *J Toxicol Environmen Health B* 10: 3-39.
63. Burke H, Leonardi-Bee J, Hashim A, et al. (2012) Prenatal and Passive Smoke Exposure and Incidence of Asthma and Wheeze: Systematic Review and Meta-analysis. *Pediatrics* 129: 735-744.
64. Chambers DM, Ocariz JM, McGuirk MF, et al. (2011) Impact of cigarette smoking on Volatile Organic Compound (VOC) blood levels in the U.S. Population: NHANES 2003-2004. *Environ Int* 37: 1321-1328.
65. Gilbert CR, Arum SM, Smith CM (2009) Vitamin D deficiency and chronic lung disease. *Can Respir J: Journal of the Canadian Thoracic Society* 16: 75.
66. Mansbach JM, Ginde AA, Camargo CA (2009) Serum 25-Hydroxyvitamin D Levels Among US Children Aged 1 to 11 Years: Do Children Need More Vitamin D *Pediatrics* 124: 1404-1410.
67. Zittermann A, Dembinski J, Stehle P (2004) Low vitamin D status is associated with low cord blood levels of the immunosuppressive cytokine interleukin-10. *Pediatric Allergy Immunol* 15: 242-246.
68. Pfeffer PE, Hawrylowicz CM (2012) Vitamin D and lung disease. *Thorax*.
69. Bozzetto S, Carraro S, Giordano G, et al. (2012) Asthma, allergy and respiratory infections: the vitamin D hypothesis. *Allergy* 67: 10-17.
70. Kozlowska E, Krzystyniak K, Drela N, et al. (1996) Thymus-directed immunotoxicity of airborne dust particles from Upper Silesia (Poland) under acute extrapulmonary studies in mice. *J Toxicol Environmen Health* 49: 563-579.
71. Busse W, Banks-Schlegel S, Noel P, et al. (2004) Future Research Directions in Asthma. *Am J Resp Crit Care* 170: 683-690.
72. Holt PG, Macaubas C, Stumbles PA, et al. (1999) The role of allergy in the development of asthma. *Nature* 402: B12-17.
73. Herberth G, Heinrich J, R□der S, et al. (2010) Reduced IFN-γ- and enhanced IL-4-producing CD4̄ cord blood T cells are associated with a higher risk for atopic dermatitis during the first 2 yr of life. *Pediatric Allergy Immunol* 21: 5-13.
74. Herberth G, Hinz D, Roder S, et al. (2011) Maternal immune status in pregnancy is related to offspring's immune responses and atopy risk. *Allergy* 66: 1065-1074.
75. Lehmann I, Thoelke A, Rehwagen M, et al. (2002) The influence of maternal exposure to volatile organic compounds on the cytokine secretion profile of neonatal T cells. *Environ Toxicol* 17: 203-210.
76. Lehmann I, Thoelke A, Weiss M, et al. (2002) T cell reactivity in neonates from an East and a West German city – results of the LISA study. *Allergy* 57: 129-136.
77. Koike Y, Hisada T, Utsugi M, et al. (2007) Glutathione redox regulates airway hyperresponsiveness and airway inflammation in mice. *Am J Respir Cell Mol Biol* 37: 322-329.
78. Kuipers H, Lambrecht BN (2004) The interplay of dendritic cells, Th2 cells and regulatory T cells in asthma. *Curr Opin Immunol* 16: 702-708.
79. Holgate ST, Davies DE, Powell RM, et al. (2007) Local genetic and environmental factors in asthma disease pathogenesis: chronicity and persistence mechanisms. 29: 793-803.

80. Ober C, Hoffjan S (2006) Asthma genetics 2006: the long and winding road to gene discovery. *Genes Immun* 7: 95-100.
81. Bilbo SD (2013) programming of neuroendocrine function by early-life experience: a critical role for the immune system. *Horm Behav* 63: 684-691.
82. Mandal M, Donnelly R, Elkabes S, et al. (2013) Maternal immune stimulation during pregnancy shapes the immunological phenotype of offspring. *Brain Behav Immun* 33: 33-45.
83. Bilbo SD, Schwarz JM (2012) The immune system and developmental programming of brain and behavior. *Front Neuroendocrinol* 33: 267-286.
84. Cory-Slechta DA, Virgolini MB, Rossi-George A, et al. (2008) Lifetime consequences of combined maternal lead and stress. *Basic Clin Pharmacol Toxicol* 102: 218-227.
85. Rappaport SM (2011) Implications of the exposome for exposure science. *J Expo Sci Environ Epidemiol* 21: 5-9.
86. Lewis R, Demmelmair H, Gaillard R, et al. (2013) The placental exposome: placental determinants of fetal adiposity and postnatal body composition. *Ann Nutr Metab* 63: 208-215.
87. Miller GW, Jones DP (2013) The nature of nurture: refining the definition of the exposome. *Toxicol Sci*: kft251.
88. Vineis P, Veldhoven K, Chadeau-Hyam M, et al. (2013) Advancing the application of omics-based biomarkers in environmental epidemiology. *Environ Mol Mutagen* 54: 461-467.
89. Nicholson JK, Wilson ID (2003) Understanding'global'systems biology: metabonomics and the continuum of metabolism. *Nat Rev Drug Discov* 2: 668-676.
90. Thacher JD, Gruzieva O, Pershagen G, et al. (2014) Pre- and Postnatal Exposure to Parental Smoking and Allergic Disease Through Adolescence. *Pediatrics* 134: 428-434.
91. Patel MM, Quinn JW, Jung KH, et al. (2011) Traffic density and stationary sources of air pollution associated with wheeze, asthma, and immunoglobulin E from birth to age 5 years among New York City children. *Environ Res* 111: 1222-1229.
92. Janssen B, Godderis L, Pieters N, et al. (2013) Placental DNA hypomethylation in association with particulate air pollution in early life. *Part Fibre Toxicol* 10: 22.
93. Janssen B, Munters E, Pieters N, et al. (2012) Placental Mitochondrial DNA Content and Particulate Air Pollution during in Utero Life. *Environ Health Perspect* 120: 1346-1352.
94. Lu L-JW, Anderson LM, Jones AB, et al. (1993) Persistence, gestation stage-dependent formation and interrelationship of benzo[a]pyrene-induced DNA adducts in mothers, placentae and fetuses of Erythrocebus patas monkeys. *Carcinogenesis* 14: 1805-1813.
95. Bradding P, Walls AF, Holgate ST (2006) The role of the mast cell in the pathophysiology of asthma. *J Allergy Clin Immunol* 117: 1277-1284.
96. Brightling CE, Bradding P, Symon FA, et al. (2002) Mast-cell infiltration of airway smooth muscle in asthma. *N Engl J Med* 346: 1699-1705.
97. Peachell P (2005) Targeting the mast cell in asthma. *Curr opin pharmacol* 5: 251-256.
98. Kleeberger SR, Ohtsuka Y, Zhang L-Y, et al. (2001) Airway responses to chronic ozone exposure are partially mediated through mast cells. *J Appl Physiol* 90: 713-723.
99. Koren HS, Hatch GE, Graham DE (1990) Nasal lavage as a tool in assessing acute inflammation in response to inhaled pollutants. *Toxicology* 60: 15-25.
100. Schierhorn K, Zhang M, Matthias C, et al. (1999) Influence of ozone and nitrogen dioxide on histamine and interleukin formation in a human nasal mucosa culture system. *Am J Respir Cell Mol Biol* 20: 1013-1019.

101. Shields RL, Gold WM (1987) Effect of inhaled ozone on lung histamine in conscious guinea pigs. *Environ Res* 42: 435-445.
102. Stenfors N, Pourazar J, Blomberg A, et al. (2002) Effect of ozone on bronchial mucosal inflammation in asthmatic and healthy subjects. *Respir Med* 96: 352-358.
103. Stenfors N, Bosson J, Helleday R, et al. (2010) Ozone exposure enhances mast-cell inflammation in asthmatic airways despite inhaled corticosteroid therapy. *Inhal Toxicol* 22: 133-139.
104. Vagaggini B, Taccola M, Conti I, et al. (2001) Budesonide reduces neutrophilic but not functional airway response to ozone in mild asthmatics. *Am J Resp Crit Care* 164: 2172-2176.
105. Janssen BG, Munters E, Pieters N, et al. (2012) Placental mitochondrial DNA content and particulate air pollution during in utero life. *Environ Health Perspect* 120: 1346-1352.
106. Prahalad A, Manchester D, Hsu I, et al. (1999) Human placental microsomal activation and DNA adduction by air pollutants. *B Environ Contam Tox* 62: 93-100.
107. Rocha e Silva IR, Lichtenfels AJF, Amador Pereira LA, et al. (2008) Effects of ambient levels of air pollution generated by traffic on birth and placental weights in mice. *Fertil Steril* 90: 1921-1924.
108. Topinka J, Binkova B, Mračková G, et al. (1997) DNA adducts in human placenta as related to air pollution and to GSTM1 genotype. *Mutatio Res-Gen- Tox En* 390: 59-68.
109. Veras MM, Damaceno-Rodrigues NR, Caldini EG, et al. (2008) Particulate urban air pollution affects the functional morphology of mouse placenta. *Biol Reprod* 79: 578-584.
110. Fujimoto A, Tsukue N, Watanabe M, et al. (2005) Diesel exhaust affects immunological action in the placentas of mice. *Environ Toxicol* 20: 431-440.
111. Menzies F, Shepherd M, Nibbs R, et al. (2010) The role of mast cells and their mediators in reproduction, pregnancy and labour. *Hum reprod update*: dmq053.
112. Woidacki K, Jensen F, Zenclussen AC (2013) Mast cells as novel mediators of reproductive processes. *Front Immunol* 4.
113. Szewczyk G, Pyzlak M, Klimkiewicz J, et al. (2012) Mast cells and histamine: do they influence placental vascular network and development in preeclampsia *Mediators Inflamm* 2012.
114. Dadvand P, Figueras F, Basagana X, et al. (2013) Ambient air pollution and preeclampsia: a spatiotemporal analysis. *Environ Health Perspect* 121: 1365-1371.
115. Lee PC, Roberts JM, Catov JM, et al. (2013) First trimester exposure to ambient air pollution, pregnancy complications and adverse birth outcomes in Allegheny County, PA. *Matern Child Health J* 17: 545-555.
116. Pereira G, Haggar F, Shand AW, et al. (2012) Association between pre-eclampsia and locally derived traffic-related air pollution: a retrospective cohort study. *J Epidemiol Commun H*
117. Woo Y, Jeong D, Chung DH, et al. (2014) The Roles of Innate Lymphoid Cells in the Development of Asthma. *Immune network* 14: 171-181.
118. Vroman H, van den Blink B, Kool M (2014) Mode of dendritic cell activation; the decisive hand in Th2 Th17 cell differentiation. Implications in asthma severity *Immunobiology*.
119. Murata Y, Shimamura T, Hamuro J (2002) The polarization of Th1 Th2 balance is dependent on the intracellular thiol redox status of macrophages due to the distinctive cytokine production. *Inter Immunol* 14: 201-212.
120. Peterson JD, Herzenberg LA, Vasquez K, et al. (1998) Glutathione levels in antigen-presenting cells modulate Th1 versus Th2 response patterns. *PNAS* 95: 3071-3076.

121. Tuzova M, Jean J-C, Hughey RP, et al. (2014) Inhibiting lung lining fluid glutathione metabolism with GGsTop as a novel treatment for asthma. *Front Pharmacol* 5.
122. Perzanowski MS, Miller RL, Tang D, et al. (2010) Prenatal acetaminophen exposure and risk of wheeze at age 5 years in an urban low-income cohort. *Thorax* 65: 118-123.
123. Penn AL, Rouse RL, Horohov DW, et al. (2007) In utero exposure to environmental tobacco smoke potentiates adult responses to allergen in BALB c mice. *Environ health perspect*: 548-555.
124. Raherison C, P nard-Morand C, Moreau D, et al. (2007) In utero and childhood exposure to parental tobacco smoke, and allergies in schoolchildren. *Resp med* 101: 107-117.
125. Jedrychowski WA, Perera FP, Majewska R, et al. (2014) Separate and joint effects of tranplacental and postnatal inhalatory exposure to polycyclic aromatic hydrocarbons: Prospective birth cohort study on wheezing events. *Pediatr Pulmonol* 49: 162-172.
126. Hansen S, Str□m M, Olsen SF, et al. (2013) Maternal concentrations of persistent organochlorine pollutants and the risk of asthma in offspring: results from a prospective cohort with 20 years of follow-up.
127. Whyatt RM, Perzanowski MS, Just AC, et al. (2014) Asthma in Inner-City Children at 5-11 Years of Age and Prenatal Exposure to Phthalates: The Columbia Center for Children's Environmental Health Cohort. *Environ Health Perspect*.
128. Spanier AJ, Kahn RS, Kunselman AR, et al. (2012) Prenatal exposure to bisphenol A and child wheeze from birth to 3 years of age. *Environ health perspects* 120: 916.
129. Stockinger B, Meglio PD, Gialitakis M, et al. (2014) The Aryl Hydrocarbon Receptor: Multitasking in the Immune System. *Annu rev immunol* 32: 403-432.
130. Kr ger T, Long M, Bonefeld-J□rgensen EC (2008) Plastic components affect the activation of the aryl hydrocarbon and the androgen receptor. *Toxicology* 246: 112-123.
131. Cayrol C, Girard J-P (2014) IL-33: an alarmin cytokine with crucial roles in innate immunity, inflammation and allergy. *Curr Opin Immunol* 31: 31-37.
132. Allakhverdi Z, Comeau MR, Smith DE, et al. (2009) CD34 sup ... sup hemopoietic progenitor cells are potent effectors of allergic inflammation. *J Allergy Clinl Immunol* 123: 472-478. e471.
133. Pr fontaine D, Lajoie-Kadoch S, Foley S, et al. (2009) Increased expression of IL-33 in severe asthma: evidence of expression by airway smooth muscle cells. *J Immunol* 183: 5094-5103.
134. Forsythe P, Ennis M (1999) Adenosine, mast cells and asthma. *Inflamm Res* 48: 301-307.
135. Gao Y-d, Cao J, Li P, et al. (2014) Th2 cytokine-primed airway smooth muscle cells induce mast cell chemotaxis via secretion of ATP. *J Asthma*: 1-21.
136. Mills KH, Dungan LS, Jones SA, et al. (2013) The role of inflammasome-derived IL-1 in driving IL-17 responses. *J Leukoc Biol* 93: 489-497.
137. Besnard A-G, Togbe D, Couillin I, et al. (2012) Inflammasome–IL-1–Th17 response in allergic lung inflammation. *J Mol Cell Biol* 4: 3-10.
138. Rappaport SM (2012) Biomarkers intersect with the exposome. *Biomarkers* 17: 483-489.
139. Holmes E, Loo R, Stamler J, et al. (2008) Human metabolic phenotype diversity and its association with diet and blood pressure. *Nature* 453: 396 - 400.
140. Tsai W, Chung R (2010) Viral hepatocarcinogenesis. *Oncogene* 29: 2309-2324.
141. Nicholson JK, Holmes E, Wilson ID (2005) Gut microorganisms, mammalian metabolism and personalized health care. *Nat Rev Microb* 3: 431-438.

142. Rappaport SM, Smith MT (2010) Environment and disease risks. *Science(Washington)* 330: 460-461.
143. Smith MT, Zhang L, McHale CM, et al. (2011) Benzene, the exposome and future investigations of leukemia etiology. *Che Biol Interact* 192: 155-159.
144. Maitre L, Fthenou E, Athersuch T, et al. (2014) Urinary metabolic profiles in early pregnancy are associated with preterm birth and fetal growth restriction in the Rhea mother-child cohort study. *BMC Medicine* 12: 110.
145. Senn T, Hazen SL, Tang W (2012) Translating metabolomics to cardiovascular biomarkers. *Prog Cardiovasc Dis* 55: 70-76.
146. Yang Y, Cruickshank C, Armstrong M, et al. (2013) New sample preparation approach for mass spectrometry-based profiling of plasma results in improved coverage of metabolome. *J Chromatogr A* 1300: 217-226.
147. Kind T, Fiehn O (2010) Advances in structure elucidation of small molecules using mass spectrometry. *Bioanalyt Rev* 2: 23-60.
148. Mattarucchi E, Baraldi E, Guillou C (2012) Metabolomics applied to urine samples in childhood asthma; differentiation between asthma phenotypes and identification of relevant metabolites. *Biomed Chromatogr* 26: 89-94.
149. Ho WE, Xu Y-J, Xu F, et al. (2013) Metabolomics reveals altered metabolic pathways in experimental asthma. *Am J Respir Cell Mol Biol* 48: 204-211.
150. Fanos V, Barberini L, Antonucci R, et al. (2011) Metabolomics in neonatology and pediatrics. *Clin Biochem* 44: 452-454.
151. Griffiths WJ, Koal T, Wang Y, et al. (2010) Targeted metabolomics for biomarker discovery. *Angew Chem Int Edit* 49: 5426-5445.
152. Carraro S, Rezzi S, Reniero F, et al. (2007) Metabolomics applied to exhaled breath condensate in childhood asthma. *Am J Resp Crit Care* 175: 986-990.
153. Tan YM, Conolly R, Chang DT, et al. (2012) Computational toxicology: application in environmental chemicals. *Methods Mol Biol* 929: 9-19.
154. Caldwell JC, Evans MV, Krishnan K (2012) Cutting Edge PBPK Models and Analyses: Providing the Basis for Future Modeling Efforts and Bridges to Emerging Toxicology Paradigms. *J Toxicol* 2012: 852384.
155. Barrett JS, Della Casa Alberighi O, Laer S, et al. (2012) Physiologically based pharmacokinetic (PBPK) modeling in children. *Clin Pharmacol Ther* 92: 40-49.
156. Bjorkman S (2005) Prediction of drug disposition in infants and children by means of physiologically based pharmacokinetic (PBPK) modelling: theophylline and midazolam as model drugs. *Br J Clin Pharmacol* 59: 691-704.
157. Vinks AA (2013) The future of physiologically based pharmacokinetic modeling to predict drug exposure in pregnant women. *CPT Pharmacometrics Syst Pharmacol* 2: e33.
158. Ruiz P, Ray M, Fisher J, et al. (2011) Development of a human Physiologically Based Pharmacokinetic (PBPK) Toolkit for environmental pollutants. *Int J Mol Sci* 12: 7469-7480.
159. Hartung T, van Vliet E, Jaworska J, et al. (2012) Systems toxicology. *ALTEX* 29: 119-128.
160. Kitano H (2002) Computational systems biology. *Nature* 420: 206-210.
161. Mc Auley MT, Wilkinson DJ, Jones JJ, et al. (2012) A whole-body mathematical model of cholesterol metabolism and its age-associated dysregulation. *BMC Syst Biol* 6: 130.

162. Mc Auley MT, Proctor CJ, Corfe BM, et al. (2013) Nutrition Research and the Impact of Computational Systems Biology. *l Comput Sci Syst Biol* 6: 271-285.
163. Wittig U, Rey M, Kania R, et al. (2014) Challenges for an enzymatic reaction kinetics database. *FEBS J* 281: 572-582.
164. Gutenkunst RN, Waterfall JJ, Casey FP, et al. (2007) Universally sloppy parameter sensitivities in systems biology models. *PLoS Comput Biol* 3: 1871-1878.
165. Carbo A, Olivares-Villagomez D, Hontecillas R, et al. (2014) Systems modeling of the role of interleukin-21 in the maintenance of effector CD4 T cell responses during chronic Helicobacter pylori infection. *MBio* 5: e01243-01214.
166. Reibman J, Marmor M, Filner J, et al. (2008) Asthma is inversely associated with Helicobacter pylori status in an urban population. *PLoS One* 3: e4060.
167. Pacifico L, Osborn JF, Tromba V, et al. (2014) Helicobacter pylori infection and extragastric disorders in children: a critical update. *World J Gastroenterol* 20: 1379-1401.
168. Wilkinson DJ (2009) Stochastic modelling for quantitative description of heterogeneous biological systems. *Nat Rev Genet* 10: 122-133.
169. Jaworska J, Gabbert S, Aldenberg T (2010) Towards optimization of chemical testing under REACH: a Bayesian network approach to Integrated Testing Strategies. *Regul Toxicol Pharmacol* 57: 157-167.
170. Chaouiya C (2007) Petri net modelling of biological networks. *Brief Bioinform* 8: 210-219.
171. Hucka M, Finney A, Bornstein BJ, et al. (2004) Evolving a lingua franca and associated software infrastructure for computational systems biology: the Systems Biology Markup Language (SBML) project. *Syst Biol (Stevenage)* 1: 41-53.
172. Ankley GT, Bennett RS, Erickson RJ, et al. (2010) Adverse outcome pathways: a conceptual framework to support ecotoxicology research and risk assessment. *Environ Toxicol Chem* 29: 730-741.
173. Kimber I, Dearman RJ, Basketter DA, et al. (2014) Chemical respiratory allergy: reverse engineering an adverse outcome pathway. *Toxicology* 318: 32-39.
174. Vinken M (2013) The adverse outcome pathway concept: a pragmatic tool in toxicology. *Toxicology* 312: 158-165.
175. National Research Council (2012) Exposure Science in the 21st Century: A Vision and a Strategy. Washington, DC: The National Academies Press. 196 p.
176. Brunekreef B (2013) Exposure science, the exposome, and public health. *Environ Mol Mutagen* 54: 596-598.
177. Bornehag CG, Blomquist G, Gyntelberg F, et al. (2001) Dampness in Buildings and Health. *Indoor Air* 11: 72-86.
178. Bornehag CG, Sundell J, Bonini S, et al. (2004) Dampness in buildings as a risk factor for health effects, (EUROEXPO). A multidisciplinary review of the literature (1998-2000) on dampness and mite exposure in buildings and health effects. *Indoor Air* 14: 243-257.
179. Bornehag CG, Sundell J, Ḧ gerhed-Engman L, et al. (2005) Association between ventilation rates in 390 Swedish homes and allergic symptoms in children. 15: 275-280.

3

Wastewater engineering applications of BioIronTech process based on the biogeochemical cycle of iron bioreduction and (bio)oxidation

Volodymyr Ivanov [1], Viktor Stabnikov [1,2], Chen Hong Guo [1], Olena Stabnikova [1], Zubair Ahmed [3], In S. Kim [4], and Eng-Ban Shuy [1]

[1] School of Civil and Environmental Engineering, Nanyang Technological University, 50 Nanyang Avenue, Singapore 639798

[2] Department of Biotechnology and Microbiology, National University of Food Technologies, 68 Volodymyrska Street, Kyiv 01601, Ukraine

[3] Department of Civil Engineering, King Abdulaziz University, Jeddah 21589, Kingdom of Saudi Arabia

[4] School of Environmental Science and Engineering, Gwangju Institute of Science and Technology, 123 Cheomdan-gwagiro, Buk-gu, Gwangju 500712, Republic of Korea

Abstract: Bioreduction of Fe(III) and biooxidation of Fe(II) can be used in wastewater engineering as an innovative biotechnology BioIronTech, which is protected for commercial applications by US patent 7393452 and Singapore patent 106658 "Compositions and methods for the treatment of wastewater and other waste". The BioIronTech process comprises the following steps: 1) anoxic bacterial reduction of Fe(III), for example in iron ore powder; 2) surface renovation of iron ore particles due to the formation of dissolved Fe^{2+} ions; 3) precipitation of insoluble ferrous salts of inorganic anions (phosphate) or organic anions (phenols and organic acids); 4) (bio)oxidation of ferrous compunds with the formation of negatively, positively, or neutrally charged ferric hydroxides, which are good adsorbents of many pollutants; 5) disposal or thermal regeration of ferric (hydr)oxide. Different organic substances can be used as electron donors in bioreduction of Fe(III). Ferrous ions and fresh ferrous or ferric hydroxides that are produced after iron bioreduction and (bio)oxidation adsorb and precipitate diferent negatively charged molecules, for example chlorinated compounds of sucralose production wastewater or other halogenated organics, as well as phenols, organic acids, phosphate, and sulphide. Reject water (return liquor) from the stage of sewage sludge dewatering on municipal wastewater treatment plants represents from 10 to 50% of phosphorus load when being recycled to the aeration tank. BioIronTech process can remove/recover more than 90% of phosphorous from this reject water thus replacing the conventional process of phosphate precipitation by ferric/ferrous salts, which are 20–100 times more expensive than iron ore, which is used in BioIronTech process. BioIronTech process can remarkably improve the aerobic and anaerobic treatments of municipal and industrial wastewaters, especially anaerobic digestion of lipid- and sulphate-containing food-processing wastewater. It can also remove the recalcitrant compounds from industrial wastewater, enhance sustainability and quality of water resources,restore eutrophicated lakes due to removal of phosphate, ammonium, and pesticides from water, and recover ammonium and phosphate from municipal and food-processing wastes.

Keywords: BioIronTech process; iron ore bioreduction; recalcitrant compounds; sucralose; phosphate recovery; ammonium recovery

1. Introduction

Biogeochemical cycling of iron—which is the fourth most abundant element on Earth's crust after oxygen, silicon, and aluminum—has a great importance for the functioning of ecosystems. There are also diverse interactions between biochemical cycles of iron and other elements, first of all carbon, nitrogen, and phosphorus [1,2,3]. The most important reactions of the biogeochemical cycle of iron are bioreduction of Fe^{3+} and biooxidation of Fe^{2+}. These reactions can be used for the numerous environmental and geotechnical engineering applications. The commercial importance of these applications is protected by US patent 7393452 and Singapore patent 106658 "Compositions and methods for the treatment of wastewater and other waste". The major claim of these patents is a combination of anaerobic and aerobic processes where a valency of a redox mediator, mainly metal cation, is cyclically changed. In the BioIronTech process, which is based on these patents, a valency of Fe^{3+} in iron ore, used as a cheap electron acceptor, is changed due to anaerobic microbial reduction to Fe^{2+}, which is oxidized chemically or microbiologically under aerobic/microaerophilic conditions to Fe^{3+} using oxygen as an electron acceptor. Produced Fe^{3+} could be further either hydrolyzed and precipitated as ferric hydroxides or reacts with organic and inorganic substances forming insoluble compounds [4,5].

There are hundreds of potential practical applications of BioIronTech process in wastewater engineering. The aim of this paper is to show how the results of the study of the biogeochemical cycle of iron in the form of BioIronTech process can be used in wastewater treatment. Biogeochemical cycling of iron includes chemical or biological oxidation of Fe^{2+} to Fe^{3+} and bioreduction of Fe^{3+} to Fe^{2+}. Major natural depots of Fe(III) in nature are iron ores, iron hydroxides of wetland deposits, and iron-containing minerals of clay. Bioreduction of Fe(III) is performed anaerobically due to either non-specific reduction by hydrogen that is produced by fermenting bacteria or using specific reduction by iron-reducing bacteria. Different organic compounds, mainly organic acids, can be used as electron donors. Produced ferrous ions are involved in hydrolysis or forming chelates with organic acids but some portion of ferrous ions is remaining soluble under anaerobic conditions even at a neutral pH. Iron-reducing bacteria are present in anaerobic digester of organic wastes. However, reduction of insoluble iron compounds does not affect methanogenesis [19,20], probably, because iron-reducing bacteria oxidize fatty acids, which are the final products of acidogenic fermentation that are not used in methanogenesis. Chelates of ferrous ions with soil humic acids or organic acids that are produced during ferric bioreduction can be oxidized by neutrophilic iron-oxidizing bacteria or can be used as electron donors in bacterial anoxygenic photosynthesis [1,2,3].

The numerous applications of BioIrontech process are based on the interactions between the biogeochemical cycles of iron, carbon, phosphate, nitrogen, and sulphur, are shown in this paper.

2. Methods

Major reagent for BioIronTech process is hematite, Fe_2O_3, of iron ore (Figure 1).

35 ± 5.0 mm 17.0 ± 3.0 mm 7.6 ± 1.9 mm 2.4 ± 0.4 mm 0.6 ± 0.1 mm

Figure 1. Hematite particles of iron ore with different sizes used in laboratory experiments and industrial scale tests.

The samples of water and wastewater were tested for Total Phosphorus (TP), Total Nitrogen (TN), Total Ferrous (TF), Total Organic Carbon (TOC), ammonium, nitrite, and nitrate concentrations, pH and Oxidation-Reduction Potential (ORP), Dissolved Oxygen Concentration (DOC), temperature, Total Suspended Solids (TSS), color and turbidity. The description of the analysis methods used is given in Table 1.

Table 1. Methods used in the research (APHA, 2005).

Parameters tested	Preservation/Pretreatment steps	Methods used	Instruments used
Total phosphorus (TP)	Samples preserved at −20 °C; Persulfate digestion method	4500-P E. Ascorbic acid method	Spectrophotometer DU 640B (Beckman Coulter, USA)
Total nitrogen (TN)	Samples preserved at −20 °C; 4500-N C. Persulfate digestion method	4500-NO_3^- F. Automated cadmium reduction method	Flow Injection Analyzer, QuickChem 8000 (LaChat Instruments, USA)
Total ferrous (Fe^{2+}) concentration	Samples were acidified at point of sampling by addition of 2 mL of concentrated HCl per 100 mL sample	3500-Fe B. Phenanthroline method	Spectrophotometer, UV-1201V (Shimadzu, Kyoto, Japan)
Total Dissolved Ferrous (Fe^{2+})	Samples were filtered and then acidified at point of sampling	3500-Fe B. Phenanthroline method	Spectrophotometer, UV-1201V (Shimadzu, Kyoto, Japan)
Total suspended solid (TSS)	Samples were refrigerated at 4 °C and tested within 7 days after sampling	2540 D. Total Suspended Solids dried at 103–105 °C	Oven
Turbidity	Samples were tested at point of sampling	—	In-situ turbidimeter

Parameters tested	Preservation/Pretreatment steps	Methods used	Instruments used
Total organic carbon (TOC)	Samples were refrigerated at 4 °C and tested within 7 days after sampling	5310 B. Combustion-infrared method	TOC analyzer C-VCSH(Shimadzu, Kyoto, Japan)
Ammonium concentration, NH_4^+	Samples were filtered, refrigerated at −20 °C and tested within 2 days	4500-NH_3 H. Flow Injection Analysis	Flow Injection Analyzer, QuickChem 8000 (LaChat Instruments, USA)
Nitrate (NO_3^-) and nitrite (NO_2^-) concentration	Samples were filtered, refrigerated at −20 °C and tested within 2 days	4500-NO_3^- F. Automated cadmium reduction method	Flow Injection Analyzer QuickChem 8000 (LaChat Instruments, USA)
Colour pH Redox potential Temperature	In-situ determination	In-situ determination	Multiprobe YSI556 (Yellow Springs, OH, USA)

Other specific methods are described in the related references below. All analyses have been made in triplicates. The statistical and correlation analysis of the results were performed with Microsoft Excel programs. Values of different parameters were expressed as the mean ± standard deviation.

3. Results

3.1. Application of BioIronTech process for methanogenic fermentation of fat-containing wastewater

Lipids are one of the major organic pollutants in municipal and food-processing wastewater. Wastewaters produced from edible oil refinery, fish processing, slaughterhouse, wool scouring and dairy products industry contain a high concentration of lipids. Physico-chemical treatment can remove 90% of lipids but final biological treatment is necessary because of remaining emulsified and/or colloidal lipids. Fats are hydrolyzed to long-chain fatty acids (LCFA) and glycerol during anaerobic digestion but LCFA are inhibitors of both acidogenic fermentation and methanogenesis mainly because of their surface activity causing damage of cell membranes. Addition of dissolved ferrous/ferric salts diminished the inhibitory effect of LCFA because of the precipitation of LCFA as iron salt. Iron (II) was used to reduce the inhibition caused by long-chain fatty acids to prokaryotes involved in anaerobic digestion [6,7]. Degradation of stearic acid, one of model compound of LCFA, was improved for 10 days in the presence of divalent iron by 150% (Figure 2).

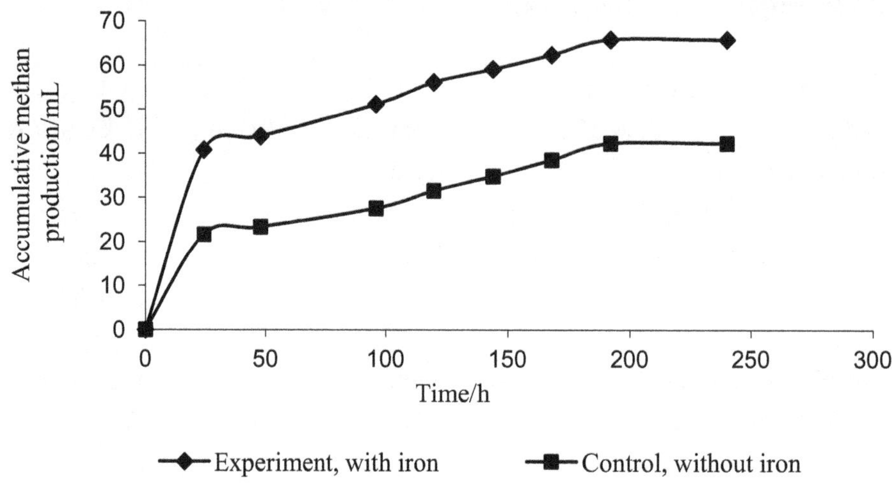

Figure 2. Accumulative methane production during fermentation of stearic acid.

The methane production rate was higher in the presence of divalent iron (0.21 mL/L/h) as compared to control (0.17 mL/L/h) where iron salt addition was absent. Methane yield was 0.1 L/g COD in experiment and 0.08 L/g COD in control.

Iron ore or even iron-containing clay can be applied for anaerobic degradation of vegetable oil. The methane production was increased 1.5 times as compared to control receiving no clay. COD removal efficiency was 98%, 80% and 77%, when iron was added to ensure the ratio of 20, 40, and 80 mg COD/mg Fe, respectively. Acetic and propionic acids were accumulated in the methanogenic reactors and inhibited methanogenesis when either iron was not present or COD/Fe ratio was higher than 20. However, no accumulation of soluble acetic and propionic acids was observed when the mass ratio of COD/Fe was below 20. So, presence of iron (II) significantly improved the anaerobic digestion of lipids [6,7]. Iron (II) can he produced in the treatment system from hematite of iron ore (Fe_2O_3), ferric hydroxide and other iron-containing minerals. These results were confirmed by other researchers [8].

3.2. Application of BioIronTech process for methanogenic fermentation of sulphate-containing wastewater

Wastewaters of fish-processing and biotechnological plants, pulp mills, chemical plants have concentration of sulphate above 1000 mg/L, which inhibits methanogenesis due to the production of toxic H_2S, competition for electron donors (acetate and hydrogen) between sulphate-reducing bacteria and methanogens, and precipitation of trace elements by sulphide. There is no production of methane when concentration of free sulphide is above about 250 mg/L. Inhibition of methanogenesis is appeared when COD/SO_4^{2-} ratio is below 10 [9].

Sulphate-reducing bacteria and iron-reducing bacteria are competitors for electron donors. Some sulphate-reducing bacteria are also able to reduce Fe(III). Therefore, an addition of ferric hydroxide and iron-reducing bacteria in anaerobic treatment of sulphate-containing wastewater inhibits sulphate bioreduction, production of sulphide, and increased removal of total organic carbon (TOC) and methane production [9,10]. Effect of ferric compound addition on anaerobic treatment of the model fish-processing wastewater depended on Fe(III) dosage, which can be determined as a molar ratio of

$Fe(III)/SO_4^{2-}$ (Table 2). These data showed efficiency of BioIronTech process for anaerobic treatment of sulphate-containing wastewater.

Table 2. The parameters of methanogenic fermentation during 10–14 days of the treatment [modified from 9,10].

Initial molar ratio of Fe^{2+}/SO_4^{2-} in the model wastewater of fish-processing plant	The rate of fermentation, g TOC/g VSS h	Final concentration of sulfide in wastewater, mg/L	Specific methane production rate, mL/gVSS day	Content of methane in biogas/% (v/v)
0.06	0.23	91	0.936	25
0.5	0.55	45	1.128	41
1.0	0.69	0	1.536	55
2.0	0.76	0	1.656	62

3.3. Application of BioIronTech process for removal of xenobiotics and recalcitrant compounds from industrial wastewaters

The removal of recalcitrant compounds from wastewater is due to (1) anoxic oxidation; (2) formation of Fe^{2+} and insoluble ferrous and ferric salts and chelates of organic acids and phenols; (3) sedimentation of polar recalcitrant molecules during coagulation of ferrous produced from iron ore.

Adsorption and precipitation of xenobiotics is due to formation of positively and negatively charged hydroxides of ferrous/ferric on anaerobic and aerobic stages of BioIronTech process:

$$Fe^{2+} + x_1H_2O \rightarrow x_2Fe(OH)^+ + x_3Fe(OH)_2 + x_4Fe(OH)_3^- \quad (1),$$

$$Fe^{3+} + x_5H_2O \rightarrow x_6Fe(OH)_2^+ + x_7Fe(OH)_3 + x_8Fe(OH)_4^- \quad (2).$$

BioIronTech process was used for the treatment of wastewater of the plant producing artificial sweetener sucralose. Sucralose (trichlorosucrose) is used in food and beverages since 1991 and now is permitted for use in over 60 countries. Sucralose is hardly biodegradable compound. It present in sewage treatment plant effluents and waterways throughout Europe and USA [11]. Sucralose and by-products of its synthesis are almost not degraded on existing biofacilities and on municipal wastewater treatment plants (MWWTP). Concentration of sucralose in effluent of MWWTP could be up to several thousand ng/L, so its concentration in surface waters in some countries is up to several hundred ng/L. Therefore, sucralose could be considered at present as a tracer of anthropogenic contamination of environment. It was shown the presence of sucralose even in North American coastal and open ocean waters [12].

The wastewater of sucralose production contains the remainder of sucralose as well as chlorinated by-products of its synthesis. These organic substances are almost not biodegradable. Their biodegradability showed as a ratio of biological and chemical consumption of oxygen for oxidation (BOD/COD ratio) is 0.003. As shown in our studies, these compounds can be partially removed from wastewater using BioIronTech process. The major mechanism of recalcitrant organics removal from wastewater of sucralose is adsorption of negatively charged chlorinated organics on positively charged particles of iron (hydr)oxides. However, without BioIronTech process an adsorption capacity of iron (hydr)oxide particles is low because of a low specific surface of the particles. However, iron-reducing bacteria in BioIrontech process renew permanently the surface of iron ore particles and produce from iron ore small particles of ferrous and ferric hydroxides with high specific surface adsorbing chlorinated organics. We showed that application of BioIronTech process with bioreduction of iron ore by added sucrose removed up to 70% of xenobiotics from the sucralose production wastewater and almost 100% of colored substances (parameter "True Color"). Therefore, BioIronTech process can significantly diminish huge quantity of granulated activated carbon that is used currently or is planned to be used for the removal of recalcitrant organics from industrial wastewater of sucralose production. This process can be designed even more effectively because iron (hydr)oxide can be regenerated by thermal treatment that will burn safely all adsorbed xenobiotics. A mechanism of the removal of xenobiotics from the wastewater of sucralose production plant is shown in Figure 3.

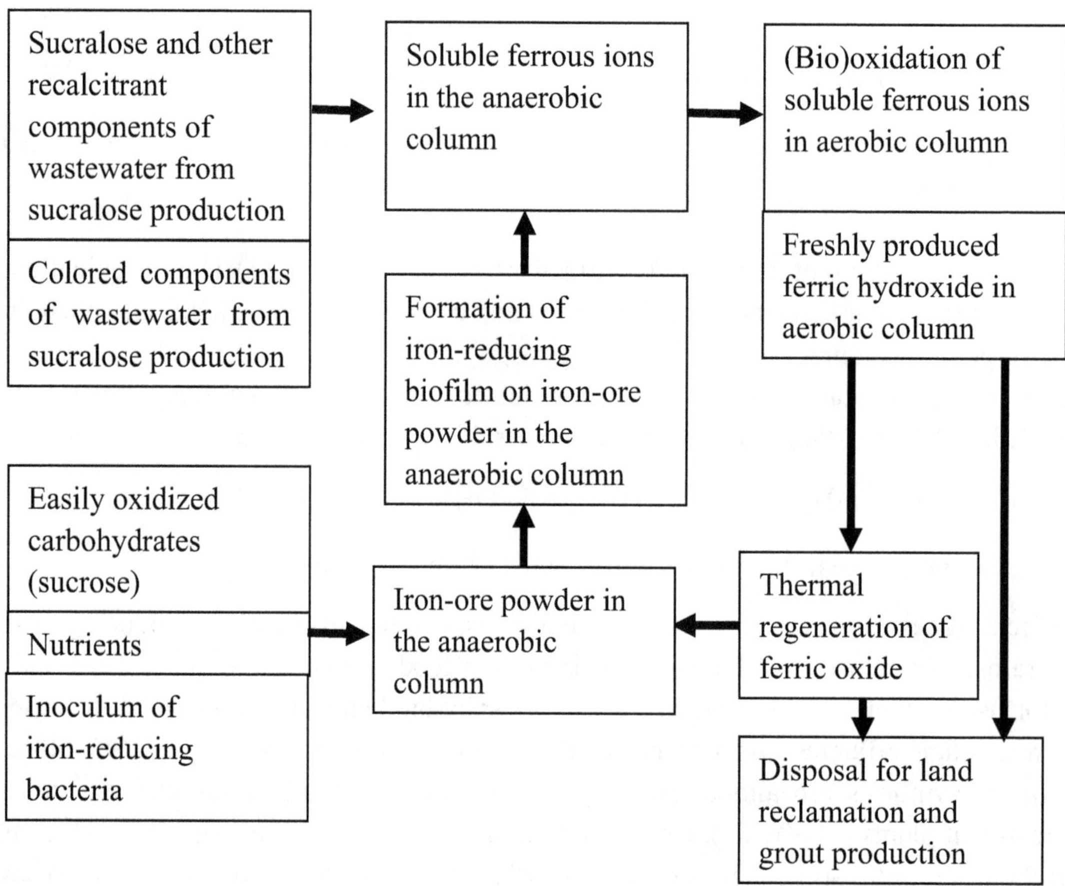

Figure 3. A mechanism of the removal of xenobiotics from the wastewater of sucralose production plant.

Fe(III) of hematite in iron ore is very cheap and industrially suitable an electron acceptor in anoxic biodegradation of xenobiotics [13,14]. For example, selected strain *Stenotrophomonas maltophilia* BK showed an ability to reduce Fe(III) using such xenobiotics as diphenylamine, m-cresol, 2,4-dichlorphenol and p-phenylphenol as sole sources of carbon under anaerobic conditions [13]. Natural estrogens such as estrone, 17β-estradiol, estriol, and the synthetic component of contraceptive pills, 17α-ethynylestradiol, enter the municipal wastewater treatment plant *via* human excretions. A significant portion of these substances remains in wastewater effluent after aerobic treatment of sewage and in returned liquor (reject water) after anaerobic digestion of activated sludge. In this study, the effect of oxidant, Fe(III), and iron-reducing bacteria on the anaerobic degradation of estrogens in returned liquor was investigated. A facultative anaerobic strain of *Alcaligenes faecalis*, capable to degrade estrogens anaerobically with ferric hydroxide or iron ore as oxidants, was isolated and tested. Biodegradation of estrogens and accumulation of their metabolites were analysed using liquid chromatography-tandem mass spectrometry. Synthetic 17α-ethynylestradiol remained resistant to anaerobic biodegradation by iron-reducing bacteria, while natural estrogens such as 17β-estradiol, estriol and estrone were removed by 92, 60 and 27%, respectively. A large fraction of 17β-estradiol was converted to estrone during the anaerobic treatment. The ability of facultative anaerobic iron-reducing bacteria to degrade estrogens can be used for the anaerobic removal of trace organics from return liquor in the municipal wastewater treatment plants [13,14].

3.4. Application of BioIronTech process for removal of phosphate from wastewater

Main point for precipitation of phosphate is production of Fe^{2+} ions (Figure 3). The rotating reactor is most suitable for the production of ferrous ions from hematite of iron ore (Table 3).

Table 3. The production of ferrous ions by iron-reducing bacteria using hematite of iron ore and reject water of the municipal wastewater treatment plant.

The sizes of iron ore particles/mm	Fe^{2+} concentration after 15 days of batch cultivation/mg/L	
	in the rotating reactor	in the static reactor
0.6	52	115
2.4	209	48
7.6	480	27

The specific rate of Fe^{2+} production negatively correlated with the size of the iron ore particles in the static reactor, which is explained by the decrease of the specific surface of the particles with increase of their sizes. Meanwhile, there was the positive correlation between the specific rates of Fe^{2+} production with the size of the iron ore particles in the rotating reactor. This strange trend was explained by experimental data of the higher concentration of the fine iron ore particles produced from iron ore particles of big size in the rotating reactor.

The reduction of ferric compounds and the production of ferrous ions led to the precipitation of phosphate. Experimental mass ratio of added Fe(III) to removed dissolved phosphate was 2 in the anaerobic digester [15] and 4 during anaerobic treatment of effluent from the anaerobic digester [7,13]. The mass ratio of removed P to produced Fe(II) in the microbiological process was 0.17 g P/g Fe^{2+} [15] and 0.22 g P/g Fe^{2+} in the chemical precipitation of phosphate by ferric salts. Addition of ferric hydroxide to the reactor of methanogenic fermentation of activated sludge followed with microbial reduction of Fe (III) and the formation of ferrous ions, which precipitated phosphate. It was shown that 66.6–99.6% of dissolved phosphate with the initial concentration of 1000–3500 mg PO_4^{3-}/L can be removed by the addition of ferric hydroxide in concentration of 6420 mg Fe (III)/L and anaerobic sludge with iron-reducing activity. Optimal ratio of added Fe(III): removed dissolved phosphate, ensured not less than 95% removal of phosphate was 2. These data could be used in a new technology of anaerobic treatment of wastewater with simultaneously removal of phosphate. An alternative technology for the phosphate precipitation from wastewater can be the production of ferrous ions using the microbial reduction of iron ore. The cost of 1 ton of Fe in iron ore is approximately US$100–150, which is much cheaper than commercial ferric or ferrous salts [16]. This biotechnology aiming to the microbial production of cheap ferrous reagent for phosphate precipitation was proposed, studied and patented [4,5,13,16,17].

3.5. Application of BioIronTech process for removal of phosphate from stormwater and the recovery of the eutrophicated ponds and lakes

Phosphate and ammonia are common pollutants of environment and their concentrations in the discharge streams are limited by 0.1–2 mg/L in almost all national environmental standards of Asian countries. Excessive phosphorus and ammonia concentrations in aquatic systems caused their eutrophication appeared in excessive growth of phototrophs, dissolved oxygen depletion, decline in recreational value of water, and appearance of foul taste of drinking water. Concentration of phosphate in the streams and collecting reservoirs is especially important because of two reasons: 1) phosphorus is triggering element in eutrophication of reservoirs, and 2) accumulation of phosphate in the bottom sediments can make pollution of reservoirs irreversible, or at least unrecoverable, for hundred years. The signs of eutrophication in the Singapore's reservoirs require urgent measures to avoid discharge of phosphate and ammonia into reservoirs. It is supposed that the main source of this pollution is urban runoff. Due to shortage of land it would be impossible to treat storm runoff in conventional artificial wetland. The enhanced engineered wetland design consists of two parts, which are a BioIronTech (BIT) reactor followed with the constructed wetland. The BIT reactor produces ferrous flux using naturally occurring local anaerobic iron-reducing bacteria to reduce iron ore and produce ferrous ions, which react with phosphate and precipitated in the deep pit before entrance to constructed wetland (Figure 4 and Figure 5). The anaerobic BioIronTech reactor for production of ferrous ions was installed as a biofilter inside the dry weather flow channel of drain (Figure 5).Effluent from anaerobic BioIronTech biofilter was mixed with the drain water (Figure 4) with the designed volume ratio of the mixing drain water and effluent of BioIronTech reactor 9:1. The goal of the construction of BioIronTech biofilter in the drain was precipitation of ferric phosphate at the settling zone at the end of the drain to prevent release of phosphate into the reservoir and mitigate eutrophication there. The length of BioIronTech biofilter, filled with iron ore, was 130 m. The monitoring of BioIronTech biofilter was continued for 8 months. Using the monitoring data, three months were considered as the maturation period needed for the formation of biofilm on iron ore particles.

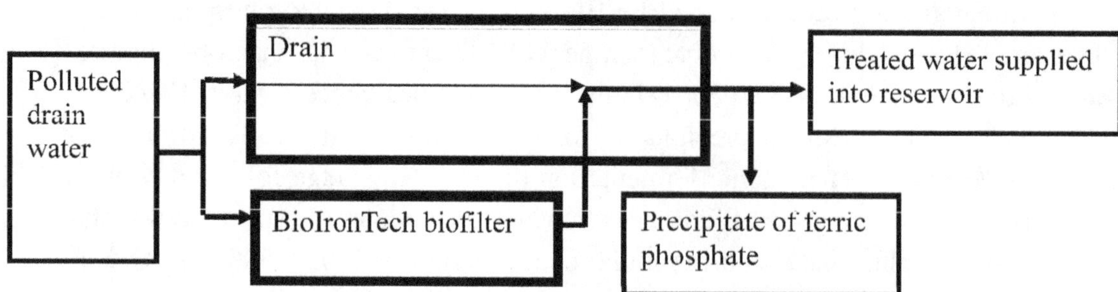

Figure 4. Flows in the continuous BioIronTech system for the treatment of polluted drain water.

Figure 5. Construction of BioIronTech biofilter in the drain of polluted stormwater.

These processes reduce the phosphate concentration in water, which enters reservoir. The average total phosphate concentration in upstream drain water was 0.24 ± 0.03 mg P/L but after BIT reactor average total phosphate concentration was 0.09 ± 0.02 mg P/L with removal efficiency by BIT reactor of 54%.The average total nitrogen concentration in upstream drain water was 0.93 ± 0.12 mg N/L and after BIT reactor an average total nitrogen concentration was 0.76 ± 0.08 mg N/L with removal efficiency of 22%. BioIronTech biotechnology requires approximately 5 times less area than constructed wetland with the same efficiency. Phosphate is recovered from wastewater by precipitation with ferrous ions produced by iron-reduced bacteria. Ammonia is recovered from polluted water probably by co-precipitation with negatively charged ferric hydroxides of cell sheathes of iron-oxidizing bacteria. The value-added by-product of BioIronTech process is fertilizer containing slow-releasing phosphate and ammonia. It is also important that the designed time of exploitation after construction is longer than 20 years. The BioIronTech biofilter of drain water is working in Singapore for last several years removing permanently about 54% of P, 22% of N, and 95% of As from collecting rain water.

Similar BioIronTech systems with the recycle of water *a priori* can be used for the recovery of the heavily eutrophicated and polluted ponds and lakes due to removal of P, N, As, other heavy metals, radionuclides, and pesticides from recycled water. There are known experimental data showing that ferric hydroxide produced by oxidation of ferrous ions is a good adsorbent of heavy metals and radionuclides containing in polluted water. For example, precipitation of freshly produced flocs of $Fe(OH)_3$ obtained by oxidation of Fe^{2+} can remove using adsorption, inclusion, and occlusion processes up to 90% of gross-alfa and 70% of gross-beta activity of groundwater containing such radionuclides as ^{226}Ra, ^{228}Ra, and ^{238}U [18,19]. Adsorption of such heavy metals as As, Cd, Co, Ni, Pb and Zn on ferric hydroxide particles is also well studied [20].

3.6. Application of BioIronTech process for removal of ammonum from wastewater

It was shown that BioIronTech process can be used for the production of Fe^{2+} ions and their slow biooxidation in water by iron-oxidizing bacteria. This oxidation prevents nitrification due to the binding of ammonium ions with negatively charged ferric hydroxide, which was produced by neutrophilic iron-oxiding bacteria under microaerophilic conditions [21,22]. The existence of ferric hydroxide as anions $Fe(OH)_4^-$ at pH from 6 to 8 is well known. These anions could bind and precipitate positively charged ammonium ions. The removal of ammonium from solution occurred during chemical oxidation of Fe(II) [21,22]. The molar ratio ammonium/iron in sediment was 0.77, while, theoretically, this ratio in $(NH_4)Fe(OH)_4$ is 1. The difference is probably due to the big size of iron hydroxide particles. It is known precipitation of ammonia in struvite, magnesium ammonium phosphate hexahydrate ($MgNH_4PO_4 \cdot 6H_2O$), but it is performed at the pH higher than 9. Another co-precipitation can be as ammonium jarosite $(NH_4)Fe_3(SO_4)_2(OH)_6$ but according to the literature data it can be done at low pH. Co-precipitation of ammonia with negatively charged Fe(III) hydroxides formed by iron-oxidizing bacteria was proposed for recovery of ammonia in aerobic treatment of wastewater [23,24].

Reject water, which is the liquid fraction produced after dewatering of anaerobically digested activated sludge on the municipal wastewater treatment plants (MWWTPs), contributes up to 80% of the nitrogen and phosphorus loads to the MWWTP. Therefore, it was proposed to combine the removal of nitrogen from reject water using the sequential biooxidation of NH_4^+ and bioreduction of NO_3^- with precipitation of phosphate by Fe^{2+} ions produced due to bioreduction of Fe^{3+} in iron ore. Bioreduction of NO_3^- decreased Fe^{3+} bioreduction rate in reject water from 37 to 21 mg Fe^{2+}/L d due to competition between NO_3^- and Fe^{3+} for electron donors. Addition of acetate as the electron donor increased both bioreduction rates of Fe^{3+} and NO_3^- but acetate interfered with the competition between nitrate and phosphate anions reacting with ferrous cations thus decreasing efficiency of the phosphate removal from reject water. Therefore, the stages of denitrification and ferric bioreduction/phosphate precipitation must be performed sequentially [17,25]. Nitrate can be also removed by oxidation of Fe^{2+}, produced by iron-reducing bacteria [25].

4. Conclusion

Major advantage of BioIronTech process is that iron ore is a relatively cheap commodity so the large scale water or wastewater treatments will be much more economical technologies than those requiring supply of expensive chemical reagents or adsorbents.Second advantage of BioIronTech process is the diversity of the pollutants that can be removed from water or wastewater. The applicability of BioIronTech process is show in the Table 4.

It is important to note that BioIronTech process is based on activity of two physiological groups of bacteria: 1) anaerobic iron-reducing bacteria and 2) usually neutrophilic microaerophilic "iron-oxidizing" bacteria, which actually do not oxidize Fe^{2+} but just oxidize organic substances of chelating envelope of ferrous atom. A lot of pure cultures of iron-reducing bacteria can be isolated and used in BioIronTech process [4,5,26,27]. However, because of the diversity of temperature, pH, salinity, type of electrons donor, type of ferrous and ferric chelating substances, redox potential, etc. in the field, it could be more practical select the enrichment cultures of iron-reducing bacteria and to cultivate them for applications [26]. Also, if necessary, neutrophilic "iron-oxidizing" bacteria can be cultivated and used in BioIronTech applications as enrichment culture.

Table 4. Potential applications of BioIronTech process in wastewater treatment.

Industry, process	Pollutant(s), which may be removed most effectively using BioIronTech process
Anoxic treatment of reject water on municipal wastewater treatment plants	Ammonia, phosphate, sulfides
Anaerobic treatment of wastewater	Sulfides, phosphate, phenols, tannins, lipids, xenobiotics, sucralose
Landfill construction and landfill leachate treatment	Organic acids, ammonium, phosphate, sulfide, heavy metals, phenols, chlorophenols, pesticides, radionuclides
Chemical industry, industrial wastewater treatment	Organic acids, cyanide, thiocyanate, phosphate, sulfate, sulfide, heavy metals, phenols, chlorophenols, pesticides, radionuclides, sucralose and chlorinatedcarbohydrates
Pharmaceutical industry, wastewater treatment	Organic acids, cyanide, thiocyanate, phosphate, sulfate, sulfide, heavy metals, phenols, chlorophenols
Brewing industry and industrial biotechnology, wastewater treatment	Ammonium, phosphate, organic acids, sulfate, sulfides
Mining, drainage waters, polluted groundwater	Cyanide, heavy metals, sulfate, arsenic
Dairy industry,wastewater treatment	Lipids, proteins, ammonia, phosphate
Vegetable oil production, treatment of wastewater	Lipids
Fish or meat processing,canning industry, wastewater treatment	Lipids, proteins, sulfate, sulfides
Shrimp aquaculture(fresh and sea water)	Ammonium, phosphate, particles, arsenic, pesticides in recycling water
Agricultural drainage	Ammonium, phosphate, pesticides, heavy metals
Electroplating and electronics, wastewater treatment	Heavy metals, sulfates, cyanides
Nuclear power plants and nuclear processing facilities; radionuclides-polluted water	Radionuclides
Aerospace industry, drainage water treatment	toxic components of fuels

Acknowledgement

The authors acknowledge support for the studies, patenting and field testing of different applications of BioIronTech process from Nanyang Technological University, Singapore; The Enterprise Challenge Fund, Prime Minister's Office, Singapore; Agency of Science and Technology (A*STAR), Singapore; Public Utilities Board (PUB), Singapore; Iowa State University, USA; National University of Food Technologies, Kyiv, Ukraine; Taras Shevchenko National University of Kyiv, Ukraine; King Abdulaziz University, Kingdom of Saudi Arabia; Gwangju Insitute of Science and Technology, South Korea; Chulalongkorn University, Thailand; King Mongkut's University of

Technology Thonburi, Thailand.

Conflict of Interest

Authors declare no conflicts of interest in this paper.

References

1. Ivanov V, Hung YT (2010) Applications of environmental biotechnology. In: *Handbook of Environmental Engineering*. Vol.10. Environmental Biotechnology, Totowa, NJ, USA: Humana Press, Inc. 2-18.
2. Ivanov V (2010) Microbiology of environmental engineering systems. In: *Handbook of Environmental Engineering*. Vol.10. Environmental Biotechnology, Humana Press, Inc. Totowa, NJ, USA: Humana Press, Inc. 19-80.
3. Ivanov V (2010) Environmental Microbiology for Engineers. CRC Press, Taylor & Francis Group, Boca Raton.402 p.
4. Tay JH, Tay STL, Ivanov V, et al. (2006) Compositions and methods for the treatment of wastewater and other waste. Patent of Singapore 106658. Date of grant 31 October 2006, date of filing 16 April 2002.
5. Tay JH, Tay STL, Ivanov V, et al. (2008) Compositions and methods for the treatment of wastewater and other waste.US Patent 7,393,452.Date of grant July 1, 2008 Date of filing 1 April 11, 2003.
6. Ivanov V, Stabnikova EV, Stabnikov VP, et al. (2002). Effects of iron compounds on the treatment of fat-containing wastewaters. *Appl Biochem & Microbiol* 38: 255-258.
7. Zubair A, Ivanov V, Hyun SH, et al. (2001) Effect of divalent iron on methanogenic fermentation of fat-containing wastewater. *Environ Eng Res* 6: 139-146.
8. Li Z, Wrenn BA, Venosa AD (2006) Effects of ferric hydroxide on methanogenesis from lipids and long-chain fatty acids in anaerobic digestion. *Water Environ Res* 78: 522-530.
9. Stabnikov VP, Ivanov VN.(2006) The effect of various iron hydroxide concentrations on the anaerobic fermentation of sulfate-containing model wastewater. *Appl Biochem & Microbiol* 42: 284-288.
10. Ivanov V, Tay STL, Wang JY, et al. (2004) Improvement of sludge quality by iron-reducing bacteria. *J Residuals Sci Tech* 1: 165-168.
11. Kessler R (2009) Sweeteners persist in waterways. *Environ Health Persp* 117: A438-A438.
12. Mead RN, Morgan JB, Avery GB, et al. (2009) Occurrence of the artificial sweetener sucralose in coastal and marine waters of the United States. *Mar Chem* 116: 13-17.
13. Ivanov V, Stabnikov V, Zhuang WQ, et al. (2005) Phosphate removal from return liquor of municipal wastewater treatment plant using iron-reducing bacteria. *J Appl Microbiol* 98: 1152-1161.
14. Ivanov V, Lim JJW, Stabnikova O, et al. (2010) Biodegradation of estrogens by facultative anaerobic iron-reducing bacteria. *Process Biochem* 45: 284-287.
15. Stabnikov V, Tay STL, Tay JH, et al. (2004) Effect of iron hydroxide on phosphate removal during anaerobic digestion of activated sludge. *Appl Biochem & Microbiol* 40: 376-380.
16. Ivanov V, Kuang SL, Guo CH, et al. (2009) The removal of phosphorus from reject water in a

municipal wastewater treatment plant using iron ore. *J Chem Technol Biot* 84:78-82.

17. Guo CH, Stabnikov V, Ivanov V (2009) The removal of phosphate from wastewater using anoxic reduction of iron ore in the rotating reactor. *Biochem Eng J* 46: 223-226.
18. Lumiste L, Munter R, Sutt J, et al. (2012) Removal of radionuclides from Estonian groundwater using aeration,oxidation, and filtration. *P Est Acad Sci* 61: 58-64.
19. Munter R (2013) Technology for the removal of radionuclides from natural water and waste management: state of the art. *P Est Acad Sci* 62: 122-132.
20. Chiang YW, Ghyselbrecht K, Santos RM, et al. (2012) Adsorption of multi-heavy metals onto water treatment residuals: Sorption capacities and applications. *Chem Eng J* 200: 405-415.
21. Ivanov V, Sihanonth P, Menasveta P (1996) Multistage-ferrous- modified-biofiltration for removal of ammonia from aquacultural water. In: *Proceedings of the Asia-Pacific Conference on Sustainable Energy and Environmental Technology*. Singapore: World Scientific Publishing 57-63.
22. Ivanov V, Stabnikova EV, Shirokih VO (1997) Influence of ferrous oxidation on the nitrification in aqueous and soil model ecosystems. *Mikrobiologia (Moscow)* 66: 428-433.
23. Ivanov V, Wang JY, Stabnikova O, et al. (2004) Iron-mediated removal of ammonia from strong nitrogenous wastewater of food processing. *Water Sci Technol* 49: 421-431.
24. Stabnikova O, Wang JY, Ivanov V (2010) Value-added biotechnological products from organic wastes. In: *Handbook of Environmental Engineering. Vol.10.Environmental Biotechnology*, Totowa, NJ, USA: Humana Press, Inc. 343-394.
25. Guo CH, Stabnikov V, Ivanov V (2010) The removal of nitrogen and phosphorus from reject water of municipal wastewater treatment plant using ferric and nitrate bioreductions. *Bioresource Technol* 101: 3992-3999.
26. Ivanov V, Stabnikov V, Hung YT (2012) Screening and selection of microorganisms for the environmental biotechnology process. In: *Handbook of Environment and Waste Management. Air and Water Pollution Control*. World Scientific Publishing Co., Inc., 1137-1149.
27. Tay JH, Tay STL, Ivanov V, et al. (2004) Application of biotechnology for industrial waste treatment. In: *Handbook of Industrial Wastes Treatment*. 2nd edition, revised and expanded by Marcel Dekker, NY, 585-618.

4

Energy and chemical efficient nitrogen removal at a full-scale MBR water reuse facility

Jianfeng Wen , Yanjin Liu, Yunjie Tu and Mark W. LeChevallier

American Water, 1025 Laurel Oak Road, Voorhees, New Jersey 08043, USA

Abstract: With stringent wastewater discharge limits on nitrogen and phosphorus, membrane bioreactor (MBR) technology is gaining popularity for advanced wastewater treatment due to higher effluent quality and smaller footprint. However, higher energy intensity required for MBR plants and increased operational costs for nutrient removal limit wide application of the MBR technology. Conventional nitrogen removal requires intensive energy inputs and chemical addition. There are drivers to search for new technology and process control strategies to treat wastewater with lower energy and chemical demand while still producing high quality effluent. The NPXpress is a patented technology developed by American Water engineers. This technology is an ultra-low dissolved oxygen (DO) operation for wastewater treatment and is able to remove nitrogen with less oxygen requirements and reduced supplemental carbon addition in MBR plants. Jefferson Peaks Water Reuse Facility in New Jersey employs MBR technology to treat municipal wastewater and was selected for the implementation of the NPXpress technology. The technology has been proved to consistently produce a high quality reuse effluent while reducing energy consumption and supplemental carbon addition by 59% and 100%, respectively. Lab-scale kinetic studies suggested that NPXpress promoted microorganisms with higher oxygen affinity. Process modelling was used to simulate treatment performance under NPXpress conditions and develop ammonia-based aeration control strategy. The application of the ammonia-based aeration control at the plant further reduced energy consumption by additional 9% and improved treatment performance with 35% reduction in effluent total nitrogen. The overall energy savings for Jefferson Peaks was $210,000 in four years since the implementation of NPXpress. This study provided an insight in design and operation of MBR plants with NPXpress technology and ultra-low DO operations.

Keywords: aeration control; MBR; NPXpress; nutrient removal; process modelling; ultra-low DO

1. Introduction

Regulations related to the nitrogen and phosphorus concentrations in the discharged wastewater are becoming more stringent. Membrane bioreactor (MBR) technology is gaining popularity and employed as tertiary treatment to produce high quality effluent that can meet discharge limits and be beneficially reused for various applications [1,2]. Compared to conventional activated sludge systems, the operating costs of MBR systems are still quite high due to air scouring for membrane cleaning and aeration demand for carbon and nitrogen removal [3]. Conventional nitrogen removal requires substantial oxygen input for nitrification and supplemental carbon source for denitrification. There have been significant interests in the recent past in the development of new nitrogen removal processes that reduce energy and chemical requirements for conventional nitrification-denitrification processes and can be applied to optimize MBR plants. Several recent advanced approaches in nitrogen removal technology include intermittent aeration, simultaneous nitrification and denitrification (SND), and the enrichment of novel microorganisms such as anaerobic ammonium oxidation (anammox) and ammonia oxidizing archaea (AOA) [4,5]. Studies have shown that aeration energy can be saved by using an intermittent aeration process to create a so-called "swing-zone" while achieving a similar level of nitrification and denitrification efficiency compared to continuous aeration processes [6,7,8].

Simultaneous nitrification and denitrification (SND) process is a promising strategy under investigation to reduce aeration energy. It has demonstrated multiple benefits over conventional processes if operated properly, and some of these include: (i) savings in aeration energy by operating the aeration tank with low dissolved oxygen (DO) concentrations; (ii) use of less supplemental carbon source; and (iii) less sludge production. SND was found in full-scale conventional treatment systems and MBR plants [9] and a DO level of 0.5 mg/L is necessary to promote a SND process [10]. American Water's research team has developed an innovative application of ultra-low DO (less than 0.2 mg/L) operations for wastewater treatment uniquely referred to as NPXpress and successfully implemented it at several full-scale MBR plants. However, the kinetics of nitrification and denitrification in those full scale MBR plants were not thoroughly investigated. It was also found that ultra-low DO operations were challenging to maintain at a stable level due to limitations in process monitoring and control. The majority of the efforts to investigate SND have been conducted at a lab-scale or pilot-scale level [11-14]. As a result, there is a gap between lab-scale studies and full scale real world operation experience for MBR plants that are operated with low energy intensity, especially for ultra-low DO operations. In addition, kinetics of nutrient removal process under NPXpress conditions need to be further investigated as well.

Process modelling is a great tool to evaluate wastewater treatment processes and helps to develop an aeration control strategy. Wastewater treatment process simulators tie together biological, chemical, and physical process models and can be used to design, upgrade, and optimize wastewater treatment plants of all types. The dissolved oxygen half saturation coefficient (K_{DO}) is a key parameter used in wastewater treatment modelling representing oxygen affinity of nitrifiers. An elevated K_{DO} value indicates lower oxygen affinity. Typically, ammonia oxidizing bacteria (AOB) have a higher oxygen affinity than nitrite oxidizing bacteria (NOB) which leads to a lower K_{DO} value for AOB. In one wastewater process simulator (BioWin, EnvironSim Associates) the default K_{DO} values for AOB and NOB are 0.25 mg/L and 0.5 mg/L, respectively. K_{DO} values vary a lot in different treatment systems [15,16]. The ultra-low DO operations in MBR systems are not well understood and determinations of K_{DO} values in the system has a significant impact on process operations since it provides optimum DO concentrations that saves energy while not compromising nitrogen removal efficiency.

Compared to the DO aeration control, ammonia-based aeration control not only reduces energy consumption, but also decreases effluent ammonia concentrations, increases denitrification efficiency, and improves biological phosphorus (bio-P) removal [17]. Simulation of ultra-low DO operations at MBR plants is still challenging and study of ammonia-aeration control using process modelling for those MBR plants to optimize treatment process is limited. In this study, an ammonia-based aeration control strategy was developed based on process simulation and the application of the aeration control strategy at the plant resulted in additional energy savings and improvement of treatment performance.

The objectives of this study were to (i) demonstrate energy and chemical efficient NPXpress technology at a full-scale MBR plant; (ii) investigate the kinetics of nitrogen and phosphorus removal in lab-scale reactors, and (iii) develop an ammonia-based aeration control strategy for energy reduction and treatment improvement using process modelling.

2. Materials and Method

2.1. Full-scale application of NPXpress

Jefferson Peaks Water Reuse Facility (WRF) is located in Hopatcong, NJ with a design flow of 473 m^3/d. The facility uses MBR technology to treat municipal wastewater and the produced effluent is used to replenish groundwater. The treatment process is shown in Figure 1. The NPXpress process was implemented at the plant in May 2010. Samples were collected periodically from influent, anoxic tank, aeration tank, membrane tank, and effluent for measuring chemical oxygen demand (COD), ammonia nitrogen (NH_3-N), nitrite nitrogen (NO_2-N), nitrate nitrogen (NO_3-N), orthophosphate (PO_4-P), and total phosphorus (TP) concentrations using a spectrophotometer (Hach DR5000, CO) and the corresponding Hach method 8000, 10031, 8507, 10020, 8048, and 8190, respectively. A portable meter (Hach HQ40d, CO) was used to measure DO concentrations. Besides this routine monitoring of process parameters, plant energy consumption and costs were monitored in 2011 to 2014. As a comparison, energy consumption without NPXpress in 2010 was also recorded and used as baseline information.

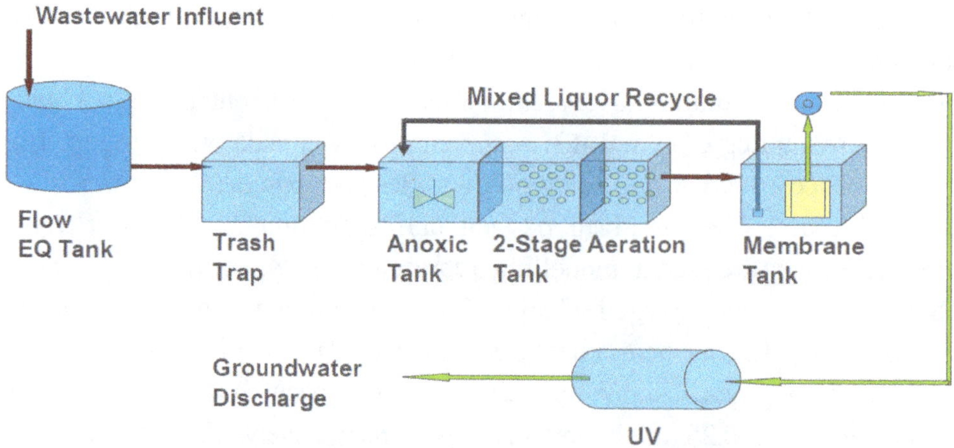

Figure 1. Treatment process layout at Jefferson Peaks WRF.

2.2. Kinetic study for biological nutrient removal

2.2.1. Biological nitrification and denitrification tests

Mixed liquor samples from Jefferson Peaks WRF were collected from the aeration tanks and membrane tanks under normal flow and load conditions. Samples were immediately transported on ice to the lab and stored in the refrigerator (4 °C) before the kinetic and characterization tests. The tests were typically started the same day or within 24 hours. Mixed liquor samples (1L) were placed in 1.2 L cylindrical glass reactors and equilibrated to room temperature (20 °C) prior to the tests. At the start of each test, 4 mL of 0.25M KH_2PO_4 buffer solution (pH = 7.2) was added to the reactors to minimize pH changes. The samples were mixed with a magnetic stirrer during the test. Typical experimental set up is shown in Figure 2. For biological nitrification tests, a stock solution (NH_3-N or NO_2-N) was added to the reactors to have a final substrate concentration of 15 mg/L. Air was pumped into the reactors through an air bubble stone. DO concentrations were well controlled with automatic DO controllers (Cole-Parmer, IL). Aliquots of 20–30 mL were collected every 15 minutes for one hour. For biological denitrification tests, nitrogen gas was initially purged into the reactors to create anaerobic conditions. Stock solutions (NH_3-N, NO_2-N, and NO_3-N) were added to the reactors to final concentrations of 15 mg/L substrates. Aliquots of 20–30 mL were collected at t = 0, 0.25, 0.75, 1.75, 2.75, and 3.75 hr. All samples from biological nitrification and denitrification tests were immediately centrifuged (Sorvall, MA) at 1500 rpm for 10 minutes, filtered using 0.45 μm membrane filters, and measured for NH_3-N, NO_2-N, and NO_3-N using Hach methods.

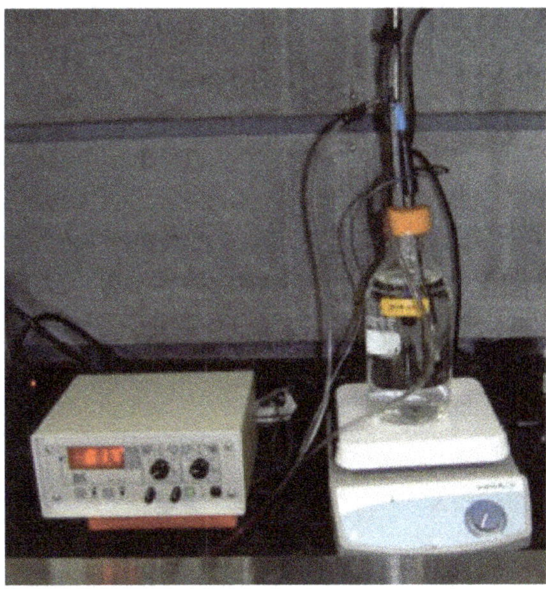

Figure 2. Kinetic study experiment set-up.

2.2.2. Determination of DO half saturation coefficient (K_{DO}) for nitrification

Biological nitrification tests were conducted under a series of DO concentrations (DO = 0.1 ± 0.05, 0.2 ± 0.05, 0.5 ± 0.05, 0.75 ± 0.05, 1 ± 0.05, and 2 ± 0.05 mg/L). The ammonia or nitrite concentrations were plotted against time for each test and linear least square regression was used to model ammonia or nitrite removal over time. The nitrification rates under each DO condition were determined by dividing the slope of the regression lines by mixed liquor volatile suspended solids (VSS) concentrations. The Monod equation (Eq. 1) was used to estimate nitrification kinetics and a linear least square method was used to calculate K_{DO} values for nitrification.

$$\mu = \mu_{max} \times \frac{DO}{K_{DO} + DO} \qquad \text{(Equation 1)}$$

Where:
μ = Specific nitrification rate for nitrifiers, mg N/L·d
μ_{max} = Maximum specific nitrification rate for nitrifiers, mg N/L·d
DO = Dissolved oxygen concentration, mg/L
K_{DO} =Dissolved oxygen half saturation coefficient, mg/L

2.2.3. Biological phosphorus release and uptake test

Phosphorus release and uptake was determined with raw wastewater influent and the mixed liquor samples in the membrane tank from the MBR plant. DO, pH, and PO_4-P were measured for both influent wastewater and mixed liquor samples before the tests. Influent wastewater (2L) and phosphate-accumulating organisms (PAOs)-containing mixed liquor samples (2L) were combined and mixed in a batch reactor, and then 0.128 g of sodium acetate was added to the reactor. The mixture was stirred continuously as to keep the sludge suspended. The biological phosphorus release and uptake test lasted 5 hours and included three sequential phases: (i) phase I, anaerobic (DO = 0 mg/L, 120 minutes); (ii) phase II, aerobic (DO = 0.1 mg/L, 60 minutes); and (iii) phase III, aerobic (DO = 4.0 mg/L, 120 minutes). Initially, the reactor was purged with nitrogen gas and kept enclosed to maintain anaerobic conditions. DO concentrations were monitored with a portable meter (Hach HQ40d, CO) and controlled by modifying air flow rate to the reactor. Samples from the batch reactor were taken every 20 or 30 minutes during the entire test to measure PO_4-P concentrations. Additionally, NH_3-N, NO_2-N and NO_3-N were measured every 60 minutes. Reactor pH was maintained between 7.0 and 8.0 during the test by adjusting with KH_2PO_4 buffer solution.

2.3. Process modelling and ammonium-based aeration control

The MBR configuration was set up in a series of three completely mixed reactors followed by membrane modules in the last reactor using BioWin 4.0 (EnviroSim Associates, Canada) to simulate the anoxic tank, aeration tank, and membrane tank. The layout of the process model is displayed in Figure 3. Modifications to the standard activated sludge model equations were necessary to account for the reduced stoichiometry of oxygen utilization and simultaneous nitrogen and phosphorus removal [18]. Typical plant operating conditions as well as default kinetic and stoichiometric parameters were initially used for the model. Both steady-state and dynamic simulations were performed to evaluate plant treatment performance. The process model was calibrated based on plant sampling data and key operational parameters such as mixed liquor suspended solids (MLSS), internal mixed liquor recycle rate, and sludge retention time (SRT). The model was further calibrated using K_{DO} values calculated from the laboratory studies. Four ammonia-based aeration control scenarios were evaluated using the calibrated model in terms of aeration and nitrogen removal efficiencies. The simulation results were used to develop aeration control strategy. YSI online sensors (Xylem, NY) were installed at the plant to monitor real time ammonia concentrations. The ammonia set-points were used to control operation of process blowers to achieve plant automation.

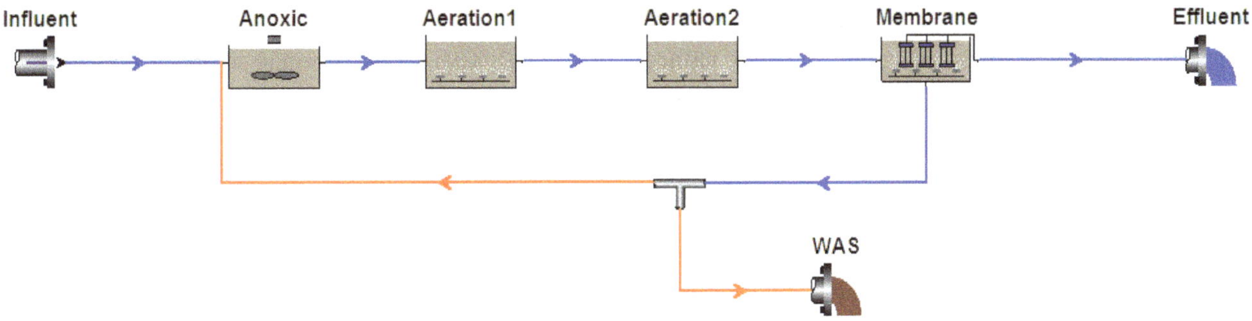

Figure 3. Process modelling layout for Jefferson Peaks WRF.

3. Results and Discussion

3.1. Treatment performance

Traditionally, it is recommended that the aeration tank has a DO level of 2.0 mg/L for optimum nitrogen removal [17,19]. Before the NPXpress conversion, the DO level in the aeration tanks of Jefferson Peaks WRF was higher than 2.0 mg/L. The NPXpress process lowered the DO level to less than 0.2 mg/L and the treatment performance was not negatively affected under ultra-low DO operations (Table 1). Specifically, effluent NH_3-N concentrations were consistently below 0.5 mg/L and the average concentration was approximately 0.18 mg/L. Similarly, the average effluent NO_3-N concentration was 4.27 mg/L and no differences were found before and after the conversion.

Table 1. Treatment Performance since NPXpress conversion[1]

	Influent (mg/L)					Effluent (mg/L)				
	COD	NH_3-N	NO_3-N	PO_4-P	TP	COD	NH_3-N	NO_3-N	PO_4-P	TP
Average	592	47.9	0.69	5.40	7.39	23	0.18	4.27	4.17	4.60
Maximum	833	89.9	1.34	8.95	9.84	58	0.47	6.96	6.50	6.36
Minimum	343	34.8	0.07	2.73	4.28	11	0.00	2.26	2.67	2.96

[1] Data are averages of 20 field measurements

The total nitrogen removal efficiency was over 90% after the conversion. The better treatment performance could be due to establishment of nitrifying and denitrifying microorganisms that can function well under ultra-low DO environment. A novel microorganism, AOA which can thrive under low DO operations, was detected in sludge samples and could lead to the NPXpress process for this MBR plant [5]. It should be noted that approximately 0.016 m^3/d supplemental carbon (Micro-C) was added to the anoxic tank for denitrification before implementing the NPXpress; while under NPXpress operations supplemental carbon addition for denitrification was completely eliminated. The total phosphorus removal efficiency was approximately 40% and PAO were accounted for biological phosphorus removal. The discussion of PAO is in the section biological phosphorus release and uptake.

Aeration accounts for half of the operational costs for wastewater treatment. The significant advantage of the NPXpress technology is to reduce oxygen demand or energy input. The overall energy consumption of Jefferson Peaks WRF was monitored before and after the NPXpress conversion. Figure 4 demonstrates energy savings after NPXpress implementation. The total electricity costs were reduced by 59% for Jefferson Peaks WRF. The NPXpress greatly reduces aeration requirements and energy costs while achieving same level of treatment performance.

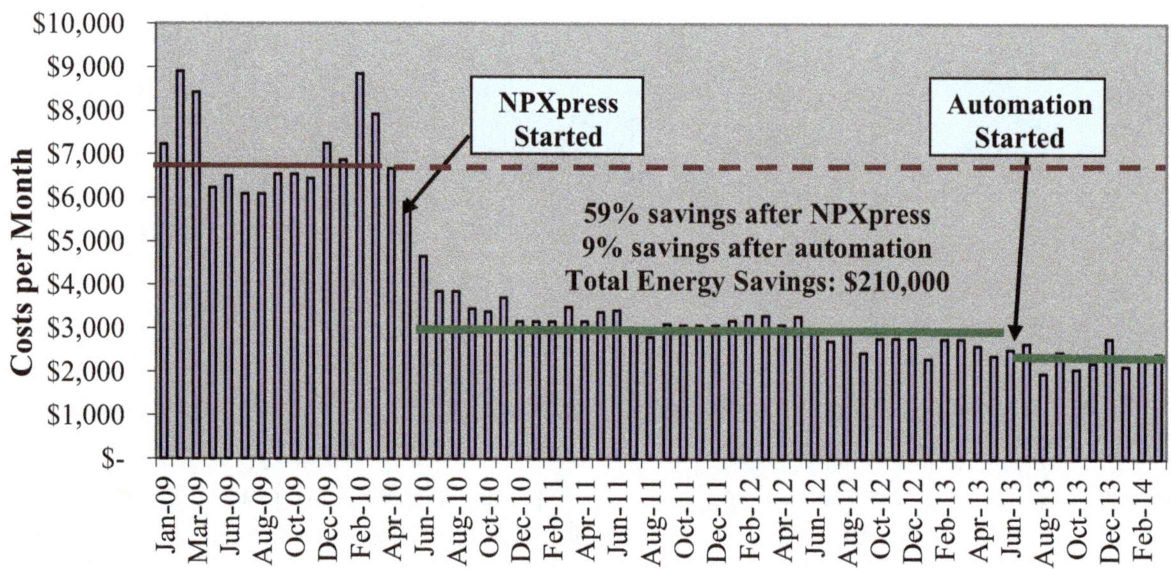

Figure 4. Overall energy costs before and after the NPXpress conversion at Jefferson Peaks WRF.

3.2. Biological nitrification and denitrification tests

Nitrification activity tests were conducted under conventional conditions (high DO) and NPXpress conditions (ultra-low DO). Figure 5 (a) displays results for a typical activity test under aerobic conditions with high DO level (DO = 2.0 mg/L). The NH_3-N concentrations decreased and correspondingly the NO_3-N concentrations increased, suggesting substantial nitrification in the reactor. The ammonia oxidation rate and nitrate production rate were 27.2 mg N/g VSS·d and 26.5 mg N/g VSS·d, respectively. It is noted that NO_2-N concentrations remained quite low in the reactor and no accumulation of nitrite was observed, indicating that all of the ammonia was initially oxidized to nitrite and the produced nitrite was then completely oxidized to nitrate. Figure 5 (b) shows nitrification results under NPXpress conditions (DO = 0.2 mg/L). The ammonia oxidation rate was 21.2 mg N/g VSS·d, which was approximately 78% of the ammonia oxidation rate under conventional conditions. The results demonstrated that microorganisms developed under NPXpress conditions had a higher affinity for oxygen. The nitrate production rate was 24.1 mg N/g VSS·d, which was higher than ammonia oxidation rate possibly due to the conversion of organic nitrogen to inorganic nitrogen during cell decay under ultra-low DO conditions. The NO_2-N concentrations remained quite low and no accumulation was observed. The aerobic activity tests under conventional and NPXpress conditions demonstrated that autotrophic nitrification was the mechanism of ammonia oxidation to nitrate.

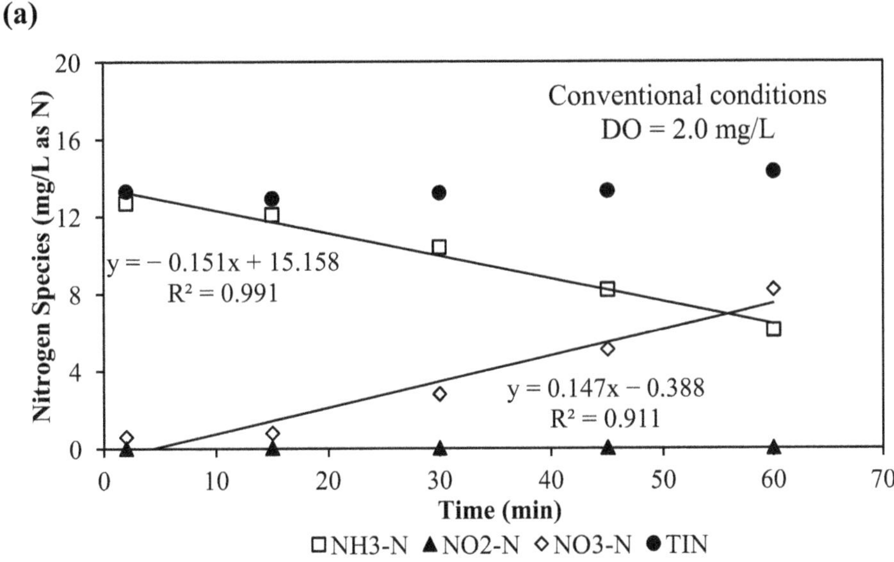

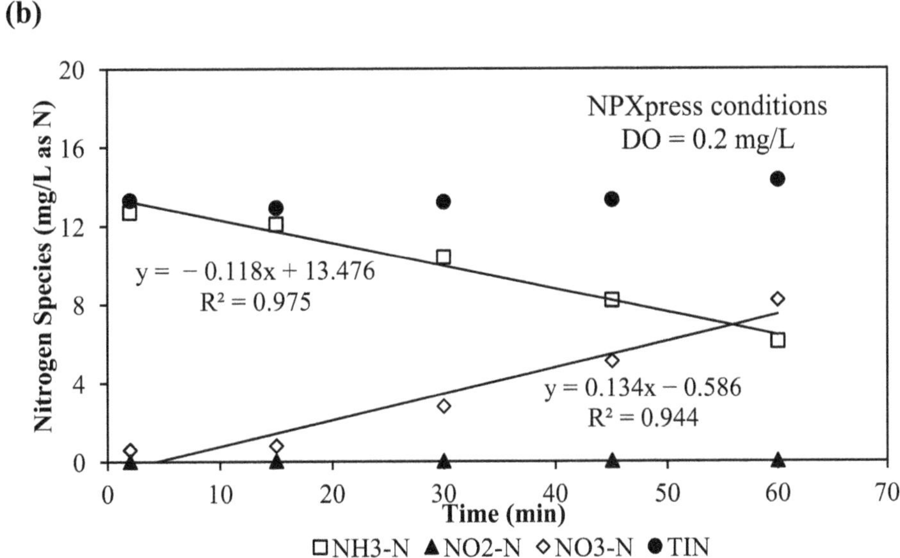

Figure 5. Biological nitrification test under conventional (a) and NPXpress (b) conditions. Note: TIN = Total inorganic nitrogen.

Figure 6 shows bacterial activity test results under anaerobic conditions. It is obvious that both NH_3-N and NO_2-N concentrations remained constant after initial conversion of ammonia to nitrite. The conversion could be due to oxygen input when injecting substrates into the reactor. The constant concentrations of NH_3-N and NO_2-N indicated that anammox was not present in the mixed liquors under ultra-low DO operations. Nitrate in the reactor decreased, indicating that under anaerobic conditions, heterotrophic denitrifying bacteria used nitrate as a terminal electron acceptor to remove nitrogen. The decrease of total inorganic nitrogen (TIN) could be attributed to endogenous respiration by heterotrophic denitrifying bacteria in the batch reactor. The results of the anaerobic activity tests suggested that traditional heterotrophic denitrification is the dominant nitrogen removal mechanism at Jefferson Peaks WRF operated under NPXpress conditions, while other novel nitrogen removal pathway (e.g. anammox) was limited.

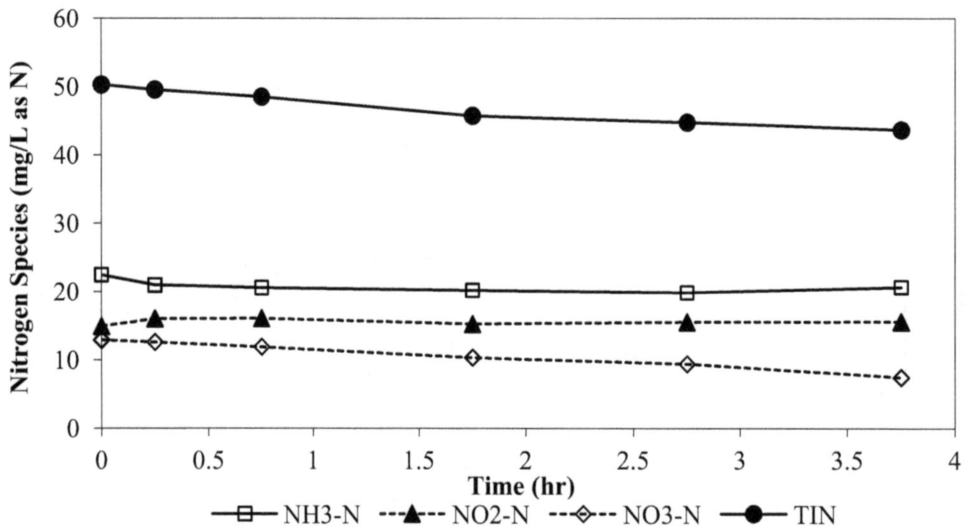

Figure 6. Biological denitrification tests under anaerobic conditions.

3.3. DO half saturation coefficient (K_{DO})

Both quantitative Polymerase Chain Reaction (qPCR) and microarray analysis demonstrated that AOB were minor participants in the nitrification process, and NOB were not detected in microarray experiments (manuscript in preparation). In addition, no anammox activities were observed as discussed before. Therefore, the dominant organisms responsible for NPXpress are still unknown and further investigations are being conducted. In this paper, ammonia oxidizing organisms (AOO) and nitrite oxidizing organisms (NOO) are used for discussions. The ammonia and nitrite oxidation rates under a series of DO conditions were calculated and the results were used to further fit the Monod equation (Eq. 1). The results of K_{DO} for AOO and NOO were 0.1 mg/L and 0.05 mg/L for mixed liquor samples from Jefferson Peaks WRF. Daebel et al. [15] reported K_{DO} ranges of 0.31–0.57 mg/L and 0.14–0.51 mg/L for AOB and NOB respectively. However, the mixed liquor samples used by Daebel et al. [15] were collected from a domestic wastewater treatment plant with DO concentrations in aeration tanks controlled between 2.5–3 mg/L. At Jefferson Peaks WRF, the DO concentrations in aeration tanks were lower than 0.2 mg/L. The results suggested that K_{DO} values were relatively low under NPXpress conditions and the NPXpress promoted microorganisms with higher oxygen affinity for biological nitrogen removal.

3.4. Biological phosphorus release and uptake

The three-phase design of the biological phosphorus release and uptake test was to mimic the actual full scale plant set-up for the anoxic zone, aeration zone (ultra-low DO), and membrane zone (high DO). Phosphorus was released under anaerobic conditions and taken up under aerated conditions (ultra-low DO and high DO; Figure 7). The rate of phosphorus uptake appeared to be faster under ultra-low DO condition than the rate under high DO condition. This indicated that DO was not the limiting factor for phosphate uptake, but the readily biodegradable carbon source might be the limiting factor for phosphate uptake in the laboratory reactor. The lab-scale results confirmed biological phosphorus removal observed at the plant.

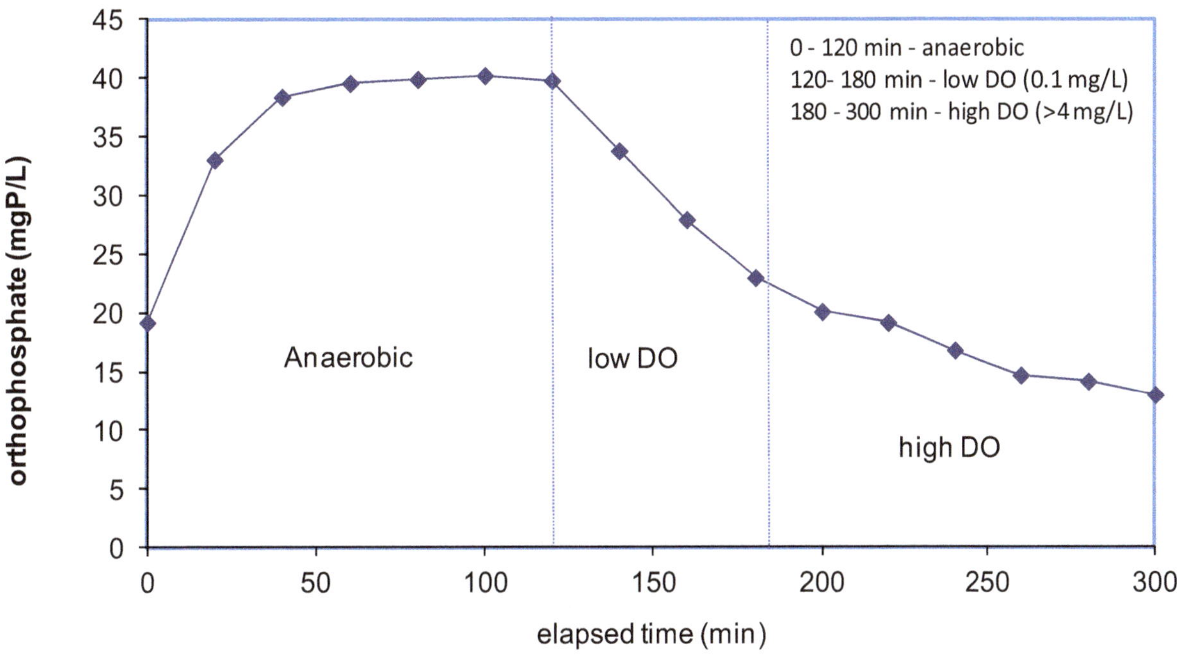

Figure 7. Biological phosphorus release and uptake in lab reactor.

3.5. Process model and ammonia-based aeration control

Once the model was calibrated, it was able to reasonably predict the plant performance for COD, NH_3-N, NO_2-N, NO_3-N and PO_4-P. BioWin Controller is a module that can augment BioWin with process control simulation. Four scenarios were set up in the BioWin Controller and simulated using the calibrated BioWin model. Figure 8 and Figure 9 show simulation results for each scenario. It is interesting that the lowest ammonia set point did not result in the highest total nitrogen removal rate. The lowest effluent total nitrogen was achieved when the set point for ammonia was between 2–3 mg/L. The reason could be attributed to unique aeration type in the biological tank. Figure 9 presents DO concentrations in aeration and membrane tanks. Higher ammonia set point led to lower DO concentrations in the membrane tank due to oxygen uptake through ammonia oxidation. Intermittent aeration was found in scenario 3 and the effluent total nitrogen concentrations were the lowest among these four scenarios. Other research teams found that intermittent aeration could improve treatment performance for nitrogen removal [6,7].

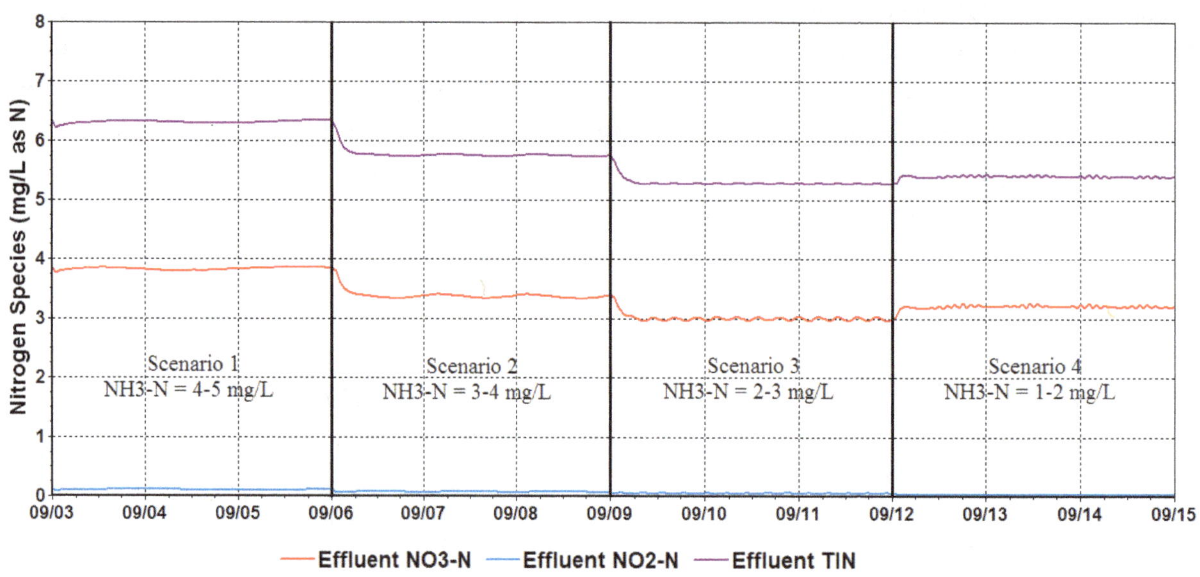

Figure 8. Simulated total effluent nitrogen with different ammonia set points.

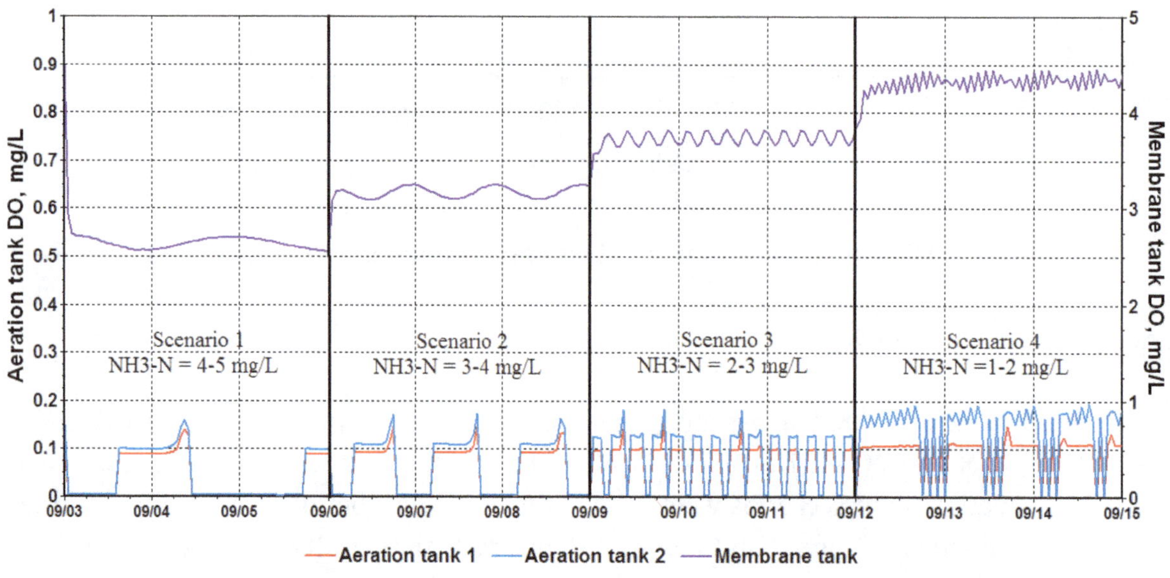

Figure 9. Simulated aeration and DO concentrations with different ammonia set points.

Based on simulation results, scenario 3 was selected as a control strategy to achieve plant automation. The ammonia concentrations in the second aeration tank were constantly monitored by online sensors. Higher ammonia concentrations will turn on the process blowers and lower concentrations will turn off the process blowers. In other words, process blowers were controller by ammonia set point similar to scenario 3. With automatic aeration control, additional 9% electricity cost savings were achieved (Figure 4). Total energy savings were over 60% and over \$210,000 in four years, which equals approximately \$0.57/m^3 operational cost savings. In addition to energy reduction, automatic aeration control improved treatment performance as well. Figure 10 shows effluent ammonia and nitrate concentrations before and after automatic aeration control. The effluent ammonia concentrations were consistently below 0.5 mg/L and effluent nitrate concentrations were gradually reduced to 2–3 mg/L. Ammonia-based aeration control reduced effluent total nitrogen by 35%, which increases treatment capacity and helps the plant prepare for more stringent discharge limits in the future.

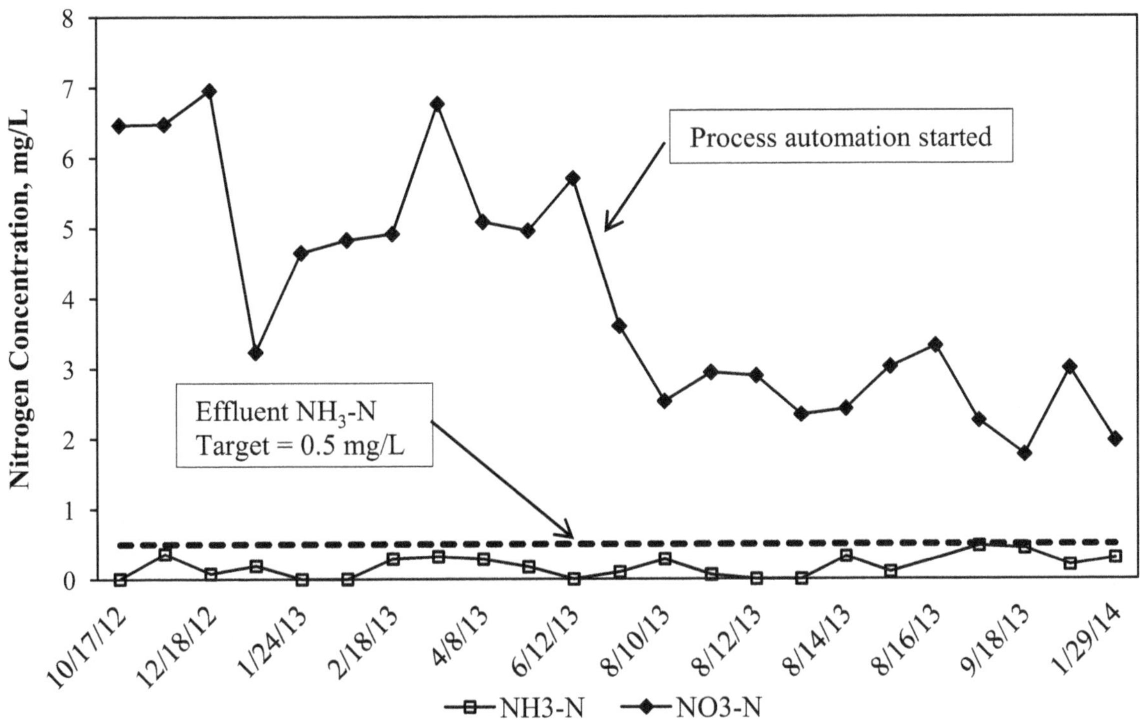

Figure 10. Treatment performance before and after process automation.

4. Conclusion

The implementation of the NPXpress technology at Jefferson Peaks WRF reduced energy consumption without compromising treatment performance. After the NPXpress conversion, the MBR plant completely eliminated the addition of supplement carbon for denitrification which reduced chemical costs, further increasing the value of NPXpress.

Lab-scale kinetic studies suggested that NPXpress promoted microorganisms with higher oxygen affinity or lower DO half saturation coefficients. The K_{DO} values for AOO and NOO were 0.1 mg/L and 0.05 mg/L respectively. The ammonia removal rate under NPXpress conditions was 78% of the rate under conventional conditions. Biological phosphorus removal was also observed in both lab-scale and full-scale studies.

Ammonia-based aeration control was developed and evaluated using process modelling. After application of automatic aeration control for NPXpress process, the overall energy savings were over 60% and approximately \$0.57/m^3 operational cost savings were achieved in four years. In addition to energy reduction, ammonia-based aeration control also improved treatment performance by reducing effluent total nitrogen by 35%.

Acknowledgments

The authors would like to thank Jim Huntington and Roger Parr for their constant support to the optimization of this full-scale MBR plant, and Dr. Patrick Jjemba for his valuable comments on an earlier version of the manuscript.

Conflict of Interest

All authors declare no conflicts of interest in this paper.

References

1. Zhou H, Smith DW (2001) Advanced technologies in water and wastewater treatment. *Can J Civil Eng* 28: 49-66.
2. Melin T, Jefferson B, Bixio D, et al. (2006) Membrane bioreactor technology for wastewater treatment and reuse. *Desalination* 187: 271-282.
3. Verrecht B, Judd SJ, Guglielmi G, et al. (2008) An aeration energy model for an immersed membrane bioreactor. *Water Res* 42: 4716-4770.
4. Schleper C, Nicol GW (2010) Ammonia-oxidising archaea – physiology, ecology and evolution. *Adv Microb Physiol* 57: 1-36.
5. Giraldo E, Jjemba P, Liu Y, et al. (2011) Ammonia oxidizing archaea, AOA, population and kinetic changes in a full scale simultaneous nitrogen and phosphorus removal MBR. *Proc Water Environ Federation* 2011: 3156-3168.
6. Huang T, Li N, Huang Y (2012) Modelling of nitrogen removal and control strategy in continuous-flow-intermittent-aeration process. *Afr J Biotechnol* 11: 10626-10631.
7. Guglielmi G, Andreottola G (2011) Alternate anoxic/aerobic operation for nitrogen removal in a membrane bioreactor for municipal wastewater treatment. *Water Sci Techn* 64: 1730-1735.
8. Curko J, Matosic M, Jakopovic HK, et al. (2010) Nitrogen removal in submerged MBR with intermittent aeration. *Desalination Water Treat* 24: 7-19.
9. Parikh C, Trivedi H, Livingston D (2011) Adecade of simultaneous nitrification and denitrification experience in over 60 conventional and MBR applications – lessons learned. *Proc Water Environ Federation* 2011: 3629-3655.
10. Holman JB, Wareham DG (2005) COD, ammonia and dissolved oxygen time profiles in the simultaneous nitrification/denitrification process. *Biochem Eng J* 22: 125-133.
11. Fu Z, Yang F, An Y, et al. (2009) Simultaneous nitrification and denitrification coupled with phosphorus removal in a modified anoxic/oxic-membrane bioreactor (A/O-MBR). *Biochem Eng J* 43: 191-196.
12. Li YZ, He YL, Ohandja DG, et al. (2008) Simutaneous nitrification-denitrification achieved by an improved internal-loop airlift MBR: comparative study. *Bioresource Technol* 99: 5867-5872.
13. Sarioglu M, Insel G, Artan N, et al. (2009) Model evaluation of simultaneous nitrification and denitrification in a membrane bioreactor operated without an anoxic reactor. *J Membrane Sci* 337: 17-27.
14. Wyffels S, Van Hulle SWH, Boeckx P, et al. (2004) Modelling and simulation of oxygen-limited partial nitrification in a membrane-assisted bioreactor (MBR). *Biotechnol Bioeng* 86: 531-542.

15. Daebel H, Manser R, Gujer W (2007) Exploring temporal variations of oxygen saturation constants of nitrifying bacteria. *Water Res* 41: 1094-1102.
16. Sharp R, Dailey S, Motyl M, et al. (2012) KDO experiments at NYC DEP's full-scale demonstration facility. *Proc Water Environ Federation* 2012: 4176-4182.
17. Rieger L, Jones RM, Dold PL, et al. (2014) Ammonia-based feedforward and feeback aeration control in activated sludge process. *Water Environ Res* 86: 63-73.
18. Henze M, Gujer W, Mino T, et al. (2000). Activated sludge models: ASM1, ASM2, ASM2d and ASM3. IWA Publishing, London.
19. Tchobanoglous G, Burton FL, Stensel HD (2004) Wastewater engineering, treatment and reuse, 4 Eds., Singapore: McGraw-Hill Education, 690.

5

Nitrate pollution of groundwater by pit latrines in developing countries

Michael R. Templeton [1], Acile S. Hammoud [1], Adrian P. Butler [1], Laura Braun [1], Julie-Anne Foucher [1], Johanna Grossmann [1], Moussa Boukari [2], Serigne Faye [3] and Jean Patrice Jourda [4]

1 Department of Civil and Environmental Engineering, Imperial College London, London, United Kingdom SW7 2AZ
2 Département des Sciences de la Terre, Université d'Abomey-Calavi, Cotonou, Bénin 01 BP 4521
3 Département de Géologie, Université Cheikh Anta Diop, Dakar, Sénégal PO Box 5005
4 Département des Sciences et Techniques de l'Eau et du Génie de l'Environnement, Université Félix Houphouët-Boigny, Abidjan, Côte d'Ivoire 22 BP 582

Abstract: Pit latrines are one of the most common forms of onsite sanitation facilities in many developing countries. These latrines are suitable as a means of isolating human waste, however, conditions within pits often lead to nitrification of the contained waste. In areas with a near-surface aquifer, the potential for nitrate pollution arising from pit latrines cannot be ignored. In this study, site visits were made to three densely populated, peri-urban areas near three West African cities (Dakar, Abidjan, Abomey-Calavi) to gather relevant information about the latrines in use and the soil and groundwater underneath the sites. Modelling was then conducted to demonstrate the potential for nitrate pollution of the groundwater from the latrines in such settings. The depth from the bottom of the pits to the water table was considered as 5, 10 or 30 m, to represent the range of aquifer depths at the study sites. Nitrate half-lives ranging from 500 to 1500 days were considered, and time scales from 6 months to several years were modelled. The results highlighted the high likelihood of nitrate pollution of groundwater reaching levels exceeding the World Health Organization guideline value for nitrate in drinking water of 50 mg/L after as short a period as two years for the aquifer situated 5 m below the pits, when considering moderate to long nitrate half-lives in the subsurface. Careful siting of latrines away from high water table areas, more frequent pit emptying, or switching to urine diversion toilets may be effective solutions to reduce nitrate passage from pit latrines into groundwater, although these solutions may not always be applicable, because of social, technical and economic constraints. The study highlights the need for more reliable data on the typical nitrate concentrations in pit latrines and the nitrate half-life in different subsurface conditions.

Keywords: nitrate; groundwater; latrine; sanitation; pollution; modelling

1. Introduction

Many low-income countries are still striving to meet their Millennium Development Goal (MDG) for access to improved sanitation, and the construction of basic pit latrines is a common strategy that is being implemented to achieve this goal. A pit latrine typically consists of a dug pit approximately 2–3 m deep and 1–2 m^2 in surface area, with the walls often lined with concrete blocks or bricks to prevent pit collapse, and with a slab and superstructure over the top of the squatting hole [1]. Lined pit walls are typically left un-mortared below the top 0.5 m [1], to allow the liquid fraction of the waste to infiltrate into the surrounding soil, leaving mainly the solid waste in the pit. However, both the liquid and solid fractions of the waste contain nitrogen, e.g. urea in urine and various forms of organic nitrogen in feces, and the biodegradation of the waste within the pit (and indeed other forms of onsite sanitation, e.g. septic tank soakaways [2,3]) often leads to nitrification of this nitrogen-containing waste, i.e. the formation of nitrate.

Nitrate is persistent and mobile in the subsurface, difficult to remove, and poses serious health concerns if it enters drinking water. The World Health Organization (WHO) has set a guideline concentration for nitrate in drinking water of 50 mg/L based on these health concerns [4]. Therefore, pit latrines may represent a serious risk to those who rely on nearby groundwater as a source of drinking water supply, especially in areas with high water tables that approach the level of the pits. Ironically therefore, regions that have met their MDG for access to improved water supply might see regression in this metric if this potential nitrate pollution arising from onsite sanitation facilities is overlooked.

This study aimed to highlight the importance of this issue by conducting modelling to represent the potential for nitrate contamination of groundwater in three low-income, peri-urban areas of West Africa with relatively high water tables. As is often the case when working in developing countries, data were limited or completely lacking for some of the key parameters of relevance to nitrate transport in the subsurface, therefore the intention of the study was not to estimate precisely the actual nitrate concentrations in the aquifers in these areas, but rather to gauge the likelihood of the nitrate levels in the affected groundwater exceeding the WHO guideline for nitrate in drinking water [4] and to estimate the time scales over which this pollution occurs.

2. Materials and Method

2.1. Site visits, data collection and estimation of parameters

Site visits were conducted to the peri-urban areas surrounding Dakar, Sénégal, Abidjan, Côte d'Ivoire, and Abomey-Calavi, Benin in the summer of 2014 to collect relevant information for the subsequent modelling. Each of these areas is characterized by high water tables, which ranged from between 5 and 30 m below the bottom of the pit latrines and are used for drinking water supply. Some of the necessary information was already available from previous hydrogeological studies which were conducted by the in-country university partners on this research [5,6], including soil hydraulic conductivity and information about depth to water table and aquifer thickness. Other information had to be collected first-hand, including an estimate of the number and dimensions of the pit latrines and the number of users per latrine. For the purposes of the subsequent modelling, a representative model sub-area was considered which was based as much as possible on the values observed through the three site visits and the previously collected information. The modelled aquifer area was 250,000 m^2, with aquifer dimensions of 1000 m in length by 250 m wide and 4 m depth. The latrine density was estimated as one latrine for every 50 m^2, and the area of each pit latrine was observed to be approximately 2 m^2, which sums to a total latrine area of 10,000 m^2. It was determined through an informal survey of the inhabitants that there was an average of 10 users per latrine. The water flux from the pit latrines was estimated as 0.002 m^3/m^2/day based on anecdotal reports of the pit usage and filling rates. The soil moisture was assumed as 0.2 and hence the pore velocity was 0.01 m/day. The longitudinal dispersivity was set at 2 m. The retardation coefficient for nitrate was assumed to be 1, reflecting the high mobility of nitrate [7,8,9]. Hydraulic conductivity data for the soil types, in the study zones, which are a mix of sand and clay, was collected from previous studies [5,6], and an average value of 7.43 m/day was used in the model. The groundwater recharge rate was estimated as being between 0.001 and 0.002 m^3/m^2/day (i.e. between 360 and 720 mm a year) and the hydraulic head gradient was estimated as ranging between 0.01 and 0.02.

There was no data for the nitrate concentration in the bottom of the pits in the study area, and there is very little data on this in the literature. However previous studies have estimated that each person excretes between 7.9–12.5 g/day of nitrogen [10-14]. To simplify the modelling, it was assumed that there was no other input of nitrate besides from the pits, e.g. none from surface runoff. A value in the middle of this range of 10 g/day per person was considered. Assuming 2.5 litres of urine per day and that 60% of the nitrogen in the waste is converted to nitrate [15], this resulted in an estimated nitrate concentration within the pit of 2400 mg/L. It should be noted that the results discussed below (and plotted in Figures 2–4) are directly scalable to this assumed value for the nitrate concentration in the bottom of the pit, therefore the results could be easily translated if a more reliable estimate of the actual concentration can be obtained, e.g. through pit sampling. Given the sensitivity of this and subsequent calculations to the value that is used for the nitrate concentration in the bottom of the pit, future studies should aim to collect data for this parameter in different settings, e.g. different regions, pit designs.

A range of nitrate half-lives have been reported in the literature, and the decay of nitrate will depend on factors such as the biological activity of the soil matrix and the nitrate sorption properties of the soil type. Since no data was available for the nitrate decay rates in the particular soils of the sites in this study, a range of half-lives were considered in the modelling to cover a range of cited values in the literature, from 500 to 1500 days [16,17].

2.2. Modelling approach

In order to provide quantitative assessment of the impact of nitrate from pit latrines on groundwater, a combined modelling approach was used. This comprised an analytical solution for steady-state reactive transport through the unsaturated (or vadose) zone, which was then used as the input to a groundwater mixing model. This approach was selected as it combines sufficient complexity to incorporate the key processes with necessary simplicity due to the lack of detailed field data. More detailed attempts have been made to simulate the problem of sludge disposal, e.g. [18,19]. However, these use the numerical model HYDRUS [20] and require detailed site data to characterize soil hydraulic properties. In contrast, a highly simplified approach has been developed by the British Geological Survey for assessing risk to groundwater from on-site sanitation. However, this does not explicitly consider the effects of denitrification and is used for local-scale assessments (i.e. at the scale of an individual borehole) [21,22].

The standard advection-dispersion-reaction model [23] (Equation 1) was therefore used to estimate the transfer of nitrate into the subsurface, from the bottom of the pit to the top of the water table, using the parameters described in section 2.1.

$$\frac{\partial C}{\partial t} = \frac{d_L v}{R}\frac{\partial^2 C}{\partial z^2} - \frac{v}{R}\frac{\partial C}{\partial z} - \frac{\lambda}{R}C \qquad (1)$$

where, C is the concentration of nitrate in unsaturated soil (g/m^3) (which is equivalent to [mg/L]), d_L is the longitudinal dispersivity (m), v is the mean pore water velocity (m/day), R is the retardation coefficient (> 1 where sorption present), λ is the linear decay coefficient (1/day), which is related to the half-life by $T\frac{1}{2} = \log(2)/\lambda$ (days).

For uniform conditions, with a nitrate concentration at the base of the pit latrine of C_0, then the concentration C_{pw} at a depth z_w, the depth of the water table below the base of the latrine, is given as:

$$C_{pw} = \frac{C_0}{2}\left[\frac{(\exp(vz_w)(1-\gamma))}{2vd_L} erfc\left(\frac{z_w - v\gamma t}{2\sqrt{vd_L t}}\right) + \frac{(\exp(vz_w)(1+\gamma))}{2vd_L} erfc\left(\frac{z_w + v\gamma t}{2\sqrt{vd_L t}}\right)\right] \qquad (2)$$

where [24]:

$$\gamma = \sqrt{1 + \frac{4\lambda d_L}{v}} \qquad (3)$$

The solution was implemented in Matlab©.

A water flow balance and nitrate mass balance were then conducted to estimate the dilution of the nitrate by the aquifer volume and the resulting overall concentration in the aquifer (C_{ao}) after different elapsed times.

Water flow balance: $Q_{ao} = Q_{ai} + A_r \cdot q_r + A_p \cdot q_p$ (4)

Nitrate mass balance: $Q_{ao} \cdot C_{ao} = Q_{ai} \cdot C_{ai} + A_r \cdot q_r \cdot C_r + A_p \cdot q_p \cdot C_{pw}$ (5)

$Q_{ai} = W \cdot H \cdot K \cdot i_i$ (6)

where Q_{ai} is the inflow into the aquifer (m^3/day), Q_{ao} is the outflow of the aquifer (m^3/day), C_{pw} is the nitrate concentration reaching the top of the water table (obtained from the advection-dispersion-reaction model, in g/m^3), C_{ao} is the nitrate concentration in the aquifer outflow (i.e. the ultimate goal of the model, in g/m^3), C_{ai} is the nitrate concentration in the aquifer inflow (assumed to be zero), C_r is the nitrate concentration from surface runoff (assumed to be zero), q_p is the water flux from each pit latrine (0.002 m/d), q_r is the groundwater recharge rate (m^3/m^2/day), A_p is the total surface area of all the pit latrines (m^2), A_r is the surface area of aquifer recharge (m^2), W is the width of the aquifer (m), H is the height of the aquifer (m), i.e. the effective mixing zone, K is the hydraulic conductivity (m/d) and i_i is the hydraulic head gradient. This is represented schematically in Figure 1.

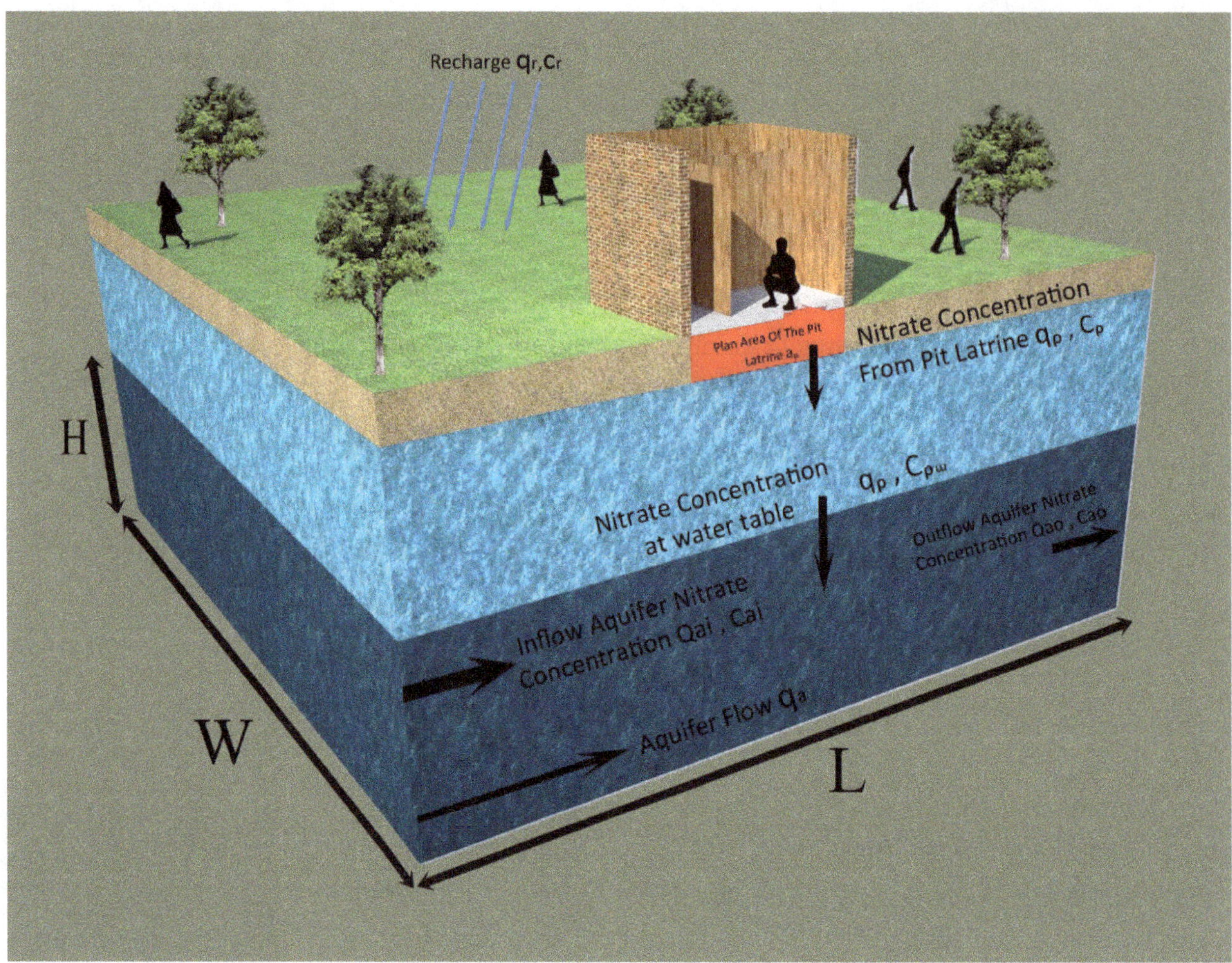

Figure 1. Schematic representation of the model parameters. The light blue zone represents the unsaturated zone and the dark blue zone represents the aquifer.

3. Results

The nitrate concentrations in the subsurface region between the bottom of the pit latrines and the water table based on the advection-dispersion-reaction model are summarized in Figures 2 to 4, considering depths from the bottom of the pits to the water table of 5, 10, and 30 m, respectively, over different time scales, ranging from 6 months up to 50 years. A range of C_{pw} values entering at the top of the aquifer, from 500 to 1500 mg/L, were then considered in the subsequent water flow and nitrate mass balance calculations, to capture the range of C_{pw} values that would occur over time under these different depth and half-life assumptions (Figures 2–4). For these flow and mass balance calculations, two hydraulic head gradients (0.01 and 0.02) and two groundwater recharge rates were considered (0.001 and 0.002 m^3/m^2/day), to represent the estimated ranges of these parameters across the study sites. The resulting C_{ao} values are tabulated below (Table 1).

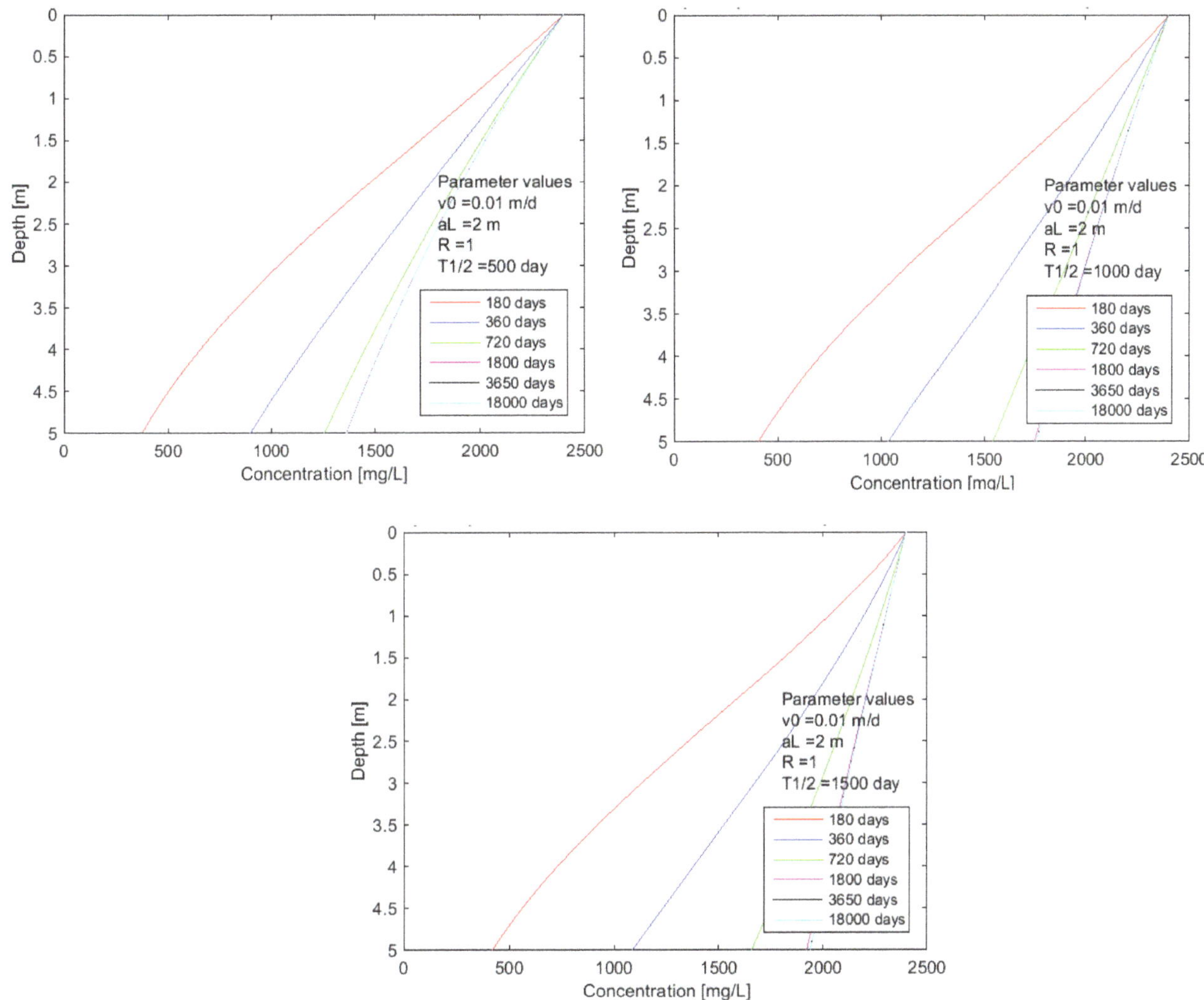

Figure 2. Depth-dependent nitrate concentrations at specified times, considering a depth from the bottom of the pits to the water table of 5 m, with nitrate half-lives assumed as 500 days (top left), 1000 days (top right), or 1500 days (bottom).

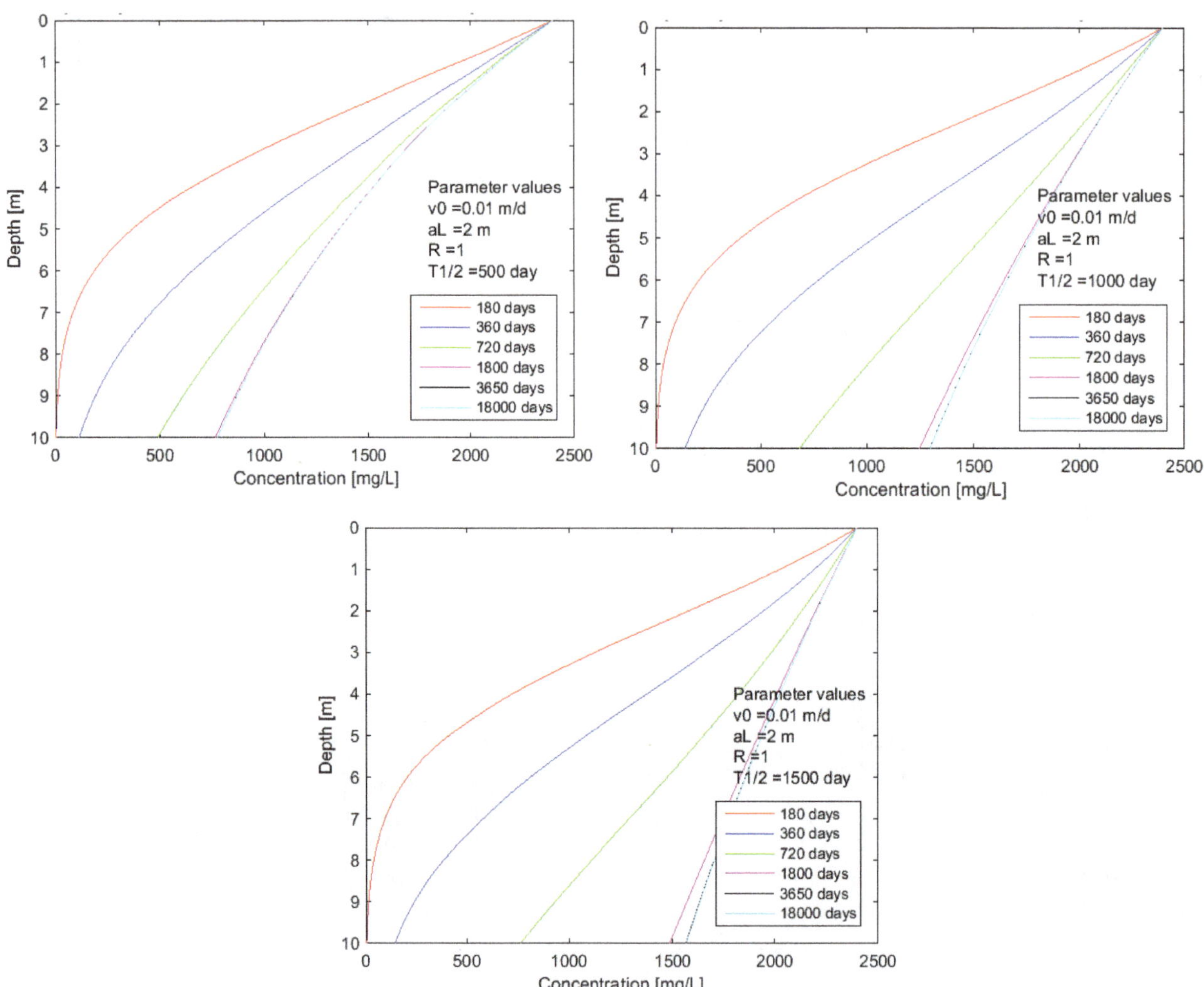

Figure 3. Depth-dependent nitrate concentrations at specified times, considering a depth from the bottom of the pits to the water table of 10 m, with nitrate half-lives assumed as 500 days (top left), 1000 days (top right), or 1500 days (bottom).

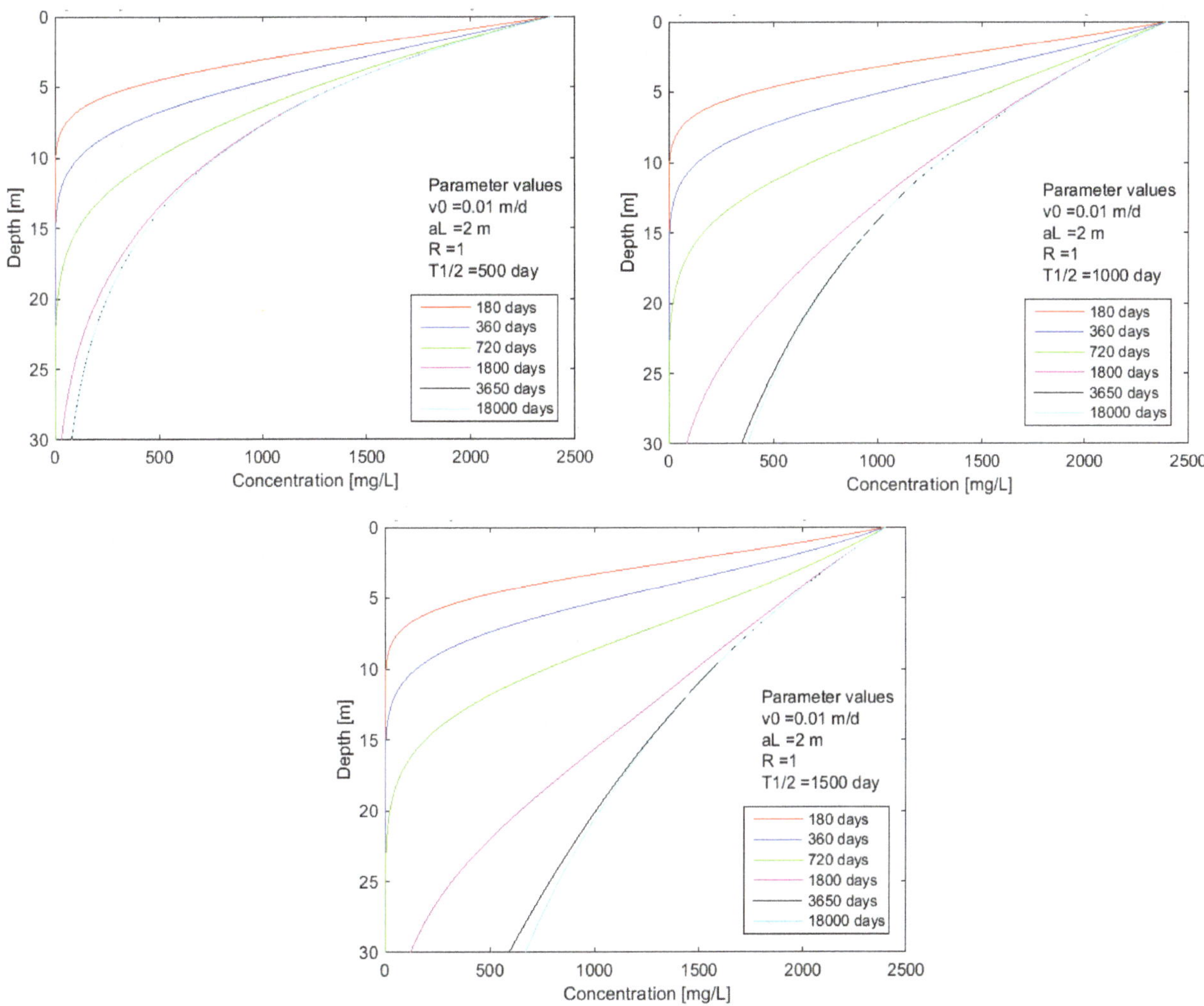

Figure 4. Depth-dependent nitrate concentrations at specified times, considering a depth from the bottom of the pits to the water table of 30 m, with nitrate half-lives assumed as 500 days (top left), 1000 days (top right), or 1500 days (bottom).

Table 1. Resulting nitrate concentrations in the aquifer (C_{ao}, mg/L), for different C_{pw} values and hydraulic head gradient (i_i) and groundwater recharge (q_r) assumptions.

C_{pw} (mg/L)	1500	1500	1250	1250	1000	1000	750	750	500	500
i_i	0.01	0.02	0.01	0.02	0.01	0.02	0.01	0.02	0.01	0.02
$q_r = 0.001$ m^3/m^2/day	87	72	73	60	58	48	44	36	29	24
$q_r = 0.002$ m^3/m^2/day	51	45	42	37	34	30	25	22	17	15

4. Discussion

Considering a 'best case' condition in this study, i.e. the maximum groundwater recharge rate (0.002 m^3/m^2/day), the highest hydraulic head gradient (0.02), and the shortest nitrate half-life (500 days), it would take over 50 years for the nitrate concentration in the aquifer to exceed the WHO guideline value for nitrate in drinking water of 50 mg/L, even for the shallowest distance from the pit down to the groundwater level of only 5 m (based on when the nitrate concentration at 5 m depth reaches 1500 mg/L in Figure 2, top left). However, the calculation is very sensitive to the nitrate half-life, and this time to exceed the WHO drinking water guideline for the 5 m depth-to-water-table case would be reduced to only two years for the moderate nitrate half-life condition of 1000 days (based on when the nitrate concentration at 5 m depth reaches 1500 mg/L in Figure 2, top right) and to even less than two years for the longest nitrate half-life condition of 1500 days (based on when the nitrate concentration at 5 m depth reaches 1500 mg/L in Figure 2, bottom). It should be noted that a commonly cited recommendation is that pit bottoms should be no less than 2 m from the groundwater level [1], while in this study the nitrate concentration penetrating to the groundwater was significant event at a depth of 5 m below the pit (except when assuming the shortest nitrate half-life). Even the 10 m depth-to-water-table case would reach the WHO drinking water guideline after approximately 5 years if the longest nitrate half-life is assumed (based on when the nitrate concentration at 10 m depth reaches 1500 mg/L in Figure 3, bottom).

The groundwater recharge rate also has a significant influence on the resulting nitrate concentrations in the aquifers, with the nitrate concentration in the aquifer increasing by approximately 60–80% when the recharge rate was halved, with all else staying the same (Table 1).

There was very little historical nitrate concentration data for the aquifers in the study locations, though relatively recent nitrate data from a well-sampling study for the Beninese study site ranged range between 0.5 to 140 mg/L [5]. The approximate ranges of contamination estimated through the modelling in this study (Table 1) fall within the field-measured range of values, which was encouraging, especially given the uncertainty regarding the concentration of nitrate in the pits and the nitrate half-life in the subsurface, as discussed earlier.

There are several potential mitigating actions that could be put in place to reduce this risk of groundwater pollution, though these may not always be suitable based on local social, technical and economic constraints. Probably the best solution is to avoid building latrines in areas with high water tables in the first place, however this is a luxury that cannot always be accommodated in booming urban areas, where the poorest people often are relegated to undesirable lands that are most prone to groundwater flooding.

For pits that are already constructed in high water table areas, more frequent emptying of the pits would at least reduce the accumulated mass of waste that leaches nitrate into the subsurface. The two years needed for the aquifer with water table 5 m below the pits to reach the WHO drinking water guideline value, explained above, may be shorter than the typical time interval between pit emptying for many households [1], for example. In that case, however, the decision on the final destination of the emptied waste [25] would also need to take careful consideration of the groundwater pollution risk, e.g. if the waste is to be dried offsite or spread on agricultural land as biosolids in areas with similarly high water tables.

Another possible solution is to change the latrine design to one that better prevents nitrate contamination. One option is to encourage the use of urine diversion toilets, which would reduce the passage of urea into the pits, however this would require significant user buy-in to be successful, both financially and from an acceptability standpoint. Also, a common recommendation for high water table areas is to build latrines on mounds rather than dug into the ground [1], although the results of the modelling in this study suggest that adding a few extra meters of depth between the pit and the groundwater level is unlikely to have a significant effect in terms of delaying nitrate penetration to the groundwater level, especially for the most shallow water table cases (e.g. water table < 5 m below the pits).

Research into novel means of achieving denitrification within a pit latrine, such as the addition nitrate-adsorbing materials to the bottom of pits or practical methods for enhancing biological denitrification in the subsurface around pits, should also be conducted. However, any such newly developed denitrification methods should be low-cost, robust to changing conditions, long-lasting, and should not require ongoing user intervention nor maintenance, which are challenging constraints.

5. Conclusions

This study has highlighted the risk posed by pit latrines to groundwater quality in terms of nitrate pollution, specifically in low-income, densely-populated peri-urban areas with only basic forms of onsite sanitation and high water tables. At a water table depth of 5 m below the pit bottom and making best case assumptions about groundwater recharge and other influencing factors, the model aquifer in this study would have reached nitrate levels approaching the WHO guideline value for nitrate in drinking water (50 mg/L) in under two years, if a conservative (i.e. moderate to long) nitrate half-life is assumed. Groundwater recharge rate was also identified as a particularly influential parameter on the resulting nitrate concentrations in the model aquifer in this study. There is a need for more data on the typical nitrate concentrations in pits and the nitrate half-lives in different subsurface conditions.

Careful siting of latrines and appropriate pit design and management are especially important in areas with high water tables. It is hoped that the modelling approach presented here may be useful for those working in other low-income regions to similarly estimate their risk of nitrate pollution and to highlight site-specific data needs for sanitation planning.

Acknowledgments

The authors acknowledge the funding support from the Royal Society and the Department for International Development, who supported this research by a grant awarded through their Africa Capacity Building Initiative (Network Grant round), entitled 'The Sustainable Onsite Sanitation (SOS) Project: Research and Capacity Building in Sanitation Engineering in West Africa', grant number AN130068. The authors also sincerely thank all the students and staff at Université Cheikh Anta Diop, Université d'Abomey-Calavi, and Université Félix Houphouët-Boigny who assisted with the site visits and data collection.

Conflict of Interest

There are no conflicts of interest to declare.

References

1. World Health Organization, Fact Sheet 3.4: Simple Pit Latrines, undated. Available from: http://www.who.int/water_sanitation_health/hygiene/emergencies/fs3_4.pdf.

2. Environment Agency, Attenuation of nitrate in the subsurface environment, 2005. Available from: https://www.gov.uk/government/uploads/system/uploads/attachment_data/file/291473/scho0605bjcs-e-e.pdf
3. Harman J, Robertson WD, Cherry JA, et al. (1996) Impacts on a sand aquifer from an old septic system: nitrate and phosphate. *Ground Water* 34: 1105-1114.
4. World Health Organization, Nitrate and nitrite in drinking-water, 2001. Available from: http://www.who.int/water_sanitation_health/dwq/chemicals/nitratenitrite2ndadd.pdf.
5. Dovonou FE (2012) *Diagonostic Qualitatif et Environnemental de l'Aquifere Superficiel du Champ de Captage Intensif de Godomey au Benin (Afrique de l'Ouest): Elements Pour un Plan d'Actions Strategiques de Protection des Resource en Eau Souterraine Exploitees*, PhD thesis, Université d'Abomey-Calavi.
6. Kouame KJ (2007) *Contribution a la Gestion Integree des Ressources en Eaux (GIRE) du District d'Abidjan (Sud de la Cote d'Ivoire): Outils d'aide a la decision pour la prevention et la protection des eaux souterraines contre la pollution*, PhD thesis, Université Félix Houphouët-Boigny.
7. Shamrukh M, Corapcioglu M, Hassona F (2001) Modelling the effect of chemical fertilizers on ground water quality in the Nile Valley Aquifer, Egypt. *Ground Water* 39: 59-67.
8. Almasri MN, Kaluarachchi JJ (2007) Modeling nitrate contamination of groundwater in agricultural watersheds. *J Hydrol* 343: 211-229
9. Lee MJ, Hwang S, Ro HM, (2014) Interpreting the effect of soil texture on transport and removal of nitrate-N in saline coastal tidal flats under steady-state flow condition. *Contin Shelf Res* 84: 35-42.
10. Schouw NL, Danteravanich S, Mosbaek H, et al. (2002) Composition of human excreta - a case study from Southern Thailand. *Sci. Tot. Environ.* 286: 155-166.
11. Del Porto D, Steinfeld C (1999) *The Composting Toilet System Book, a Practical Guide to Choosing, Planning and Maintaining composting Toilet Systems, a Water-Saving, Pollution-Preventing Alternative*, Concord, Massachussetts, The Center for Ecological Pollution Prevention.
12. Drangert JO, Bew J, Winblad U (1997) Ecological Sanitation, *Proceedings of the Sida Sanitation Workshop*, Balingsholm, Sweden, Sida.
13. Aillery M, Gollehon, N, Johanson R, et al., Managing Manure to Improve Air and Water Quality, Economic Research report 9, United States Department of Agriculture, 2005. Available from: http://www.ers.usda.gov/media/851714/err9.pdf
14. Cruz MC, Cacciabiu DG, Gil JF, et al. (2012) The impact of point source pollution on shallow groundwater used for human consumption in a threshold country. *J Environ Monit* 14: 2338-2349.
15. Kimmel GE (1984) Non-point contamination of groundwater on Long Island, New York. In: National Research Council (ed.) *Groundwater contamination.* Washington, National Academy Press, 120-126.
16. Frind E, Duynisveld W, Strebel O, et al. (1990) Modeling of multicomponent transport with microbial transformation in ground water: the Fuhrberg case. *Water Resour Res* 26: 1707-1719.
17. Herbert M, Kovar K (1998) *Groundwater Quality: Remediation and Protection.* Wallingford, International Association Hydrological Sciences.

18. Adadzi PC (2012) *Deep Row Trenching of Pit Latrine and Waste Water Treatment Works Sludge: Water and Nutrient Fluxes In Forest Plantations.* MSc Dissertation. School of Agricultural, Earth and Environmental Sciences, University of KwaZulu-Natal.
19. Heatwole KK, McCray JE (2007) Modeling potential vadose-zone transport of nitrogen from onsite wastewater systems at the development scale. *J. Contam. Hydrol.*, 91: 184-201.
20. Simunek J, van Genuchten MT, Sejna M (2006). *The HYDRUS Software Package for Simulating Two- and Three- Dimensional Movement of Water, Heat, and Multiple Solutes in Variably-Saturated Media, Technical Manual,* Version 1.0, PC Progress, Prague, Czech Republic, 241.
21. ARGOSS (2001) *Guidelines for assessing the risk to groundwater from on-site sanitation. British Geological Survey Commissioned Report*, CR/01/142. 97.
22. ARGOSS (2002) *Assessing Risk to Groundwater from On-site Sanitation: Scientific Review and Case Studies British Geological Survey Commissioned Report,* CR/02/079N. 105.
23. Butler AP, Brook C, Godley A, et al. (2003) Attenuation of landfill leachate in unsaturated sandstone. *Proceedings of the Ninth International Waste Management and Landfill Symposium, Sardinia*, October 2003.
24. van Genuchten MT, Alves WJ (1982) Analytical solution of the one-dimensional convective-dispersive solute transport equation, *Tech. Bull. U.S. Dep. Agric.*, 1661.
25. Thye YP, Templeton MR, Ali M (2011) A critical review of technologies for pit latrine emptying in developing countries. *Crit Rev Environ Sci Technol* 41: 1793-1819.

6

How synergistic or antagonistic effects may influence the mutual hazard ranking of chemicals

Lars Carlsen[1,2]

[1] Awareness Center Linkøpingvej 35, Trekroner, DK-4000 Roskilde, Denmark
[2] Center of Physical Chemical Methods of Research and Analysis, al-Farabi Kazakh National University, 96A Tole Bi st., 050012, Almaty, Kazakhstan

Abstract: The presence of various agents, including humic materials, nanomaterials, microplastics, or simply specific chemical compounds, may cause changes in the apparent persistence, bioaccumulation, and/or toxicity (PBT) of a chemical compound leading to an either increased or decreased PBT characteristics and thus an increased or decreased hazard evaluation. In the present paper, a series chloro-containing obsolete pesticides is studied as an illustrative example. Partial order methodology is used to quantify how changed P, B, or T characteristics of methoxychlor (MEC) influences the measure of the hazard of MEC, relative to the other 11 compounds in the series investigated. Not surprisingly, an increase in one of the three indicators (P, B, or T) lead to an increased average order and thus an increased relative hazard as a result of a synergistic effect. A decrease in one of the indicator values analogously causes a decreased average order/relative hazard through an antagonistic effect; the effect, however, being less pronounced. It is further seen that the effect of changing the apparent value of the three indicators is different. Thus, persistence apparently is more important that bioaccumulation which again appears more important than toxicity, which is in agreement with previous work. The results are discussed with reference to the European chemicals framework on registration, evaluation and authorization of chemicals (REACH) framework.

Keywords: synergistic effects; antagonistic effects; partial order ranking; Hasse diagram; average order; chemical hazard

---

1. Introduction

Typically, evaluation of the environmental hazard of a chemical, as may be elucidated through its persistence, bioaccumulation, and toxicity (PBT) characteristics, is performed for the 'isolated' chemical, possibly relative to others [1]. However, this procedure does not take into account the possible presence of substances or agents which one way or another may cause a change in the apparent P, B, and/or T values. Hence, it is well known that complexation may change individual values [2]. Recent and ongoing discussions on the possible use of nanomaterials as transporters for drug into the cells [3] and the role of microplastics [4] are further examples where complexation may change the inherent characteristics of organic molecules. In this respect, it should be emphasized that if sorption of drugs to nanoparticles can ease the transport of the drugs into the cell, a transport of highly unwanted substances into the cell could also be facilitated. It should be noted that the present study simply looks at the possible effects as a function of changed PBT characteristics of methoxychlor (MEC) without regard to how these changes have actually been introduced.

Little is known about such synergistic or antagonistic effects, although significant amounts of work have been devoted to the area [5]. In the present study a group of 12 "obsolete" pesticides (OPs), all banned by the Stockholm Convention [6], is studied. The mutual ranking of the 12 objects in this group of OPs has recently been studied by Sailaukhanuly *et al.* [1] based on P, B, and T data.

The objective of the study is the application of partial order methodology to elucidate and quantify how the mutual ranking of the OPs is subject to changes as a result of combination effects on the apparent P, B, or T characteristics of a single chemical, such as methoxychlor (MEC). MEC has been chosen arbitrarily, but serves as an exemplary example, as it ranks more or less in the middle when unperturbed indicators are applied [1], for example, in the absence of any combinatorial effects. To elucidate the effect, we followed the changes in rank of MEC as a function of the perturbation of P, B, or T indicators, respectively, where the mutual ranking of the compounds were a measure of the relative hazard. Obviously, the study could be further enlarged by perturbing two or three indicators simultaneously. However, this has not been included in the present paper in order not to introduce confuse.

2. Methods

2.1. Data and indicators

Quantitative Structure-Activity Relation (QSAR) generated data for the 12 obsolete pesticides (OPs), banned by the Stockholm Convention (2008), that was used as indicators. These comprise persistence (P), bioaccumulation (B), and toxicity (T) [1]. All indicators were obtained using the EPI Suite modeling tool [7]. Hence, the persistence was estimated by the BioWin module using the ultimate biodegradation as given by BDP3 of the BioWin module, the bioaccumulation by the BCF module, and the toxicity as the LC50 value for fish (LC50fish) by the EcoSAR module [7]. For the subsequent partial ordering, the persistence and the toxicity were applied as inverse values of BDP3 and LC50fish, respectively in order to assure a uniform direction (low to high) for all three indicators (Table 1).

P was estimated using the ultimate biodegradation (BDP3) as obtained through the EPI Suite's BioWin module [7,8] (Table 1). The BioWin module estimates a chemical's environmental biodegradation rate in hours, hours to days, days, days to weeks, and so on depending on the approximate amount of time needed for biodegradation to be "complete". These estimates were converted to half-lives by considering that 6 half-lives constituted "complete" or "ultimate" biodegradation and eventually to the biodegradation potential BDP. Chemicals with BDPs < 2

(1/BDP3 > 0.5) are associated with a medium persistence potential, whereas chemicals with BDPs < 1.75 (1/BDP3 > 0.57) are assigned a high persistence potential [8].

B, expressed as the logarithmic bioconcentration factor logBCF, was obtained applying the BCF module of the EPI Suite [7,8] (Table 1). BCFs were estimated from the octanol-water partition coefficient and a series of structural correction factors. Chemicals with BCFs > 1000 (log BCF > 3), but < 5000 (log BCF < 3.67) were assigned a medium bioconcentration potential [8]. Under REACH BCF > 2000 (log BCF > 3.3) are considered bioaccumulating, denoted B [9]. Chemicals with BCFs > 5000 (log BCF > 3.67) were assigned a high bioconcentration potential [8], which under REACH is denoted vB (very bioaccumulating) [9].

In general, in order to be classified as toxic within the PBT regime requires a chronic NOEC < 0.01 mg/L [9]. In the present study, the LC50 value for fish was adopted as measure of the T (Table 1). The toxicity data were estimated by the EcoSAR module of the EPI Suite [7].

Table 1. EPI Suite derived indicator values for the 12 OPs

ID	CAS No.	Trivial name	Short	Pers	BioA	Tox
				1/BDP3	logBCF	1/LC50fish
OP1	50-29-3	p,p-DDT	DDT	0.8361	4.2260	58.8235
OP2	72-55-9	p,p-DDE	DDE	0.6033	3.9620	9.8039
OP3	72-54-8	p,p-DDD	DDD	0.5758	3.6390	10.4167
OP4	72-43-5	methoxychlor	MEC	0.6611	3.0190	6.3291
OP5	309-00-2	Aldrin	ALD	1.3941	3.9560	37.0370
OP6	60-57-1	dieldrin	DIE	1.4852	3.0980	2.8169
OP7	76-44-8	heptachlor	HCL	1.9026	3.2760	6.4935
OP8	57-74-9	chlordane	CHL	3.6792	3.7710	12.8205
OP9	58-88-9	lindane (g-HCH)	LIN	0.6590	2.3990	0.4333
OP10	118-74-1	hexachlorbenzene	HCB	0.7518	3.4480	11.3636
OP11	82-68-8	pentachlor nitrobenzene	PCN	0.7441	2.7280	2.0492
OP12	87-86-5	pentachlor phenol	PCP	0.6120	3.0450	1.0060

2.2. Partial ordering

2.2.1 Overview

Partial order methodology is an alternative way to study a multi-indicator system (MIS), as it recently has been argued by Bruggemann and Carlsen [10]. Partial ordering can be considered as a non-parametric method, because in contrast to standard multidimensional analyses, no assumptions about linearity or distribution of the indicators are made nor are necessary.

2.2.2. Some technical explanations

In Partial Ordering, the only mathematical relation among the objects (in the present case, the

12 obsolete pesticides; Table 1) is "≤" [11,12]. The "≤" relation is the basis for a comparison of the OPs and constitutes a graph. An OPx is connected with an OPy if and only if the relation $y \leq x$ holds. The crucial question is: under which conditions it can be claimed that $x \leq y$? Thus, considering a system that can be described by a series of indicators—r_j, OP, OPx—and characterized by the set of indicators $r_j(OPx)$, $j = 1, ..., m$, allowing for comparison to another OP, OPy, characterized by the indicators $r_j(OPy)$, when

$$r_i(OPy) \leq r_i(OPx) \text{ for all } i = 1, \ldots, m \quad (1)$$

In the present study m equals 3 (section 2.1).

Equation 1 has a strict requirement for establishing a comparison. It demands that all indicators of OPy should be lower (or at least equal) than those of OPx through comparison of the single indicators, respectively. Note that in the present example, the lower the indicator values are, the 'better' they are considered to be, i.e., lower persistence, lower bioaccumulation and lower toxicity, respectively. Hence, let X be the group of OPs studied, i.e., $X = \{OP1, OP2, \ldots, OP12\}$, OPx will be ordered higher (worse) than OPy, i.e., OPx > OPy, if at least one of the indicator values for OPx is higher (worse) than the corresponding indicator value for OPy and no indicator for OPx is lower (better) than the corresponding indicator value for OPy. However, if $r_j(OPx) > r_j(OPy)$ for some indicator j and $r_i(OPx) < r_i(OPy)$ for some other indicator i, OPx and OPy will be called incomparable (notation: OPx ‖ OPy) expressing the mathematical contradiction due to conflicting indicator values. When all indicator values for OPx are equal to the corresponding indicator values for OPy, i.e., $r_j(OPx) = r_j(OPy)$ for all j, the two objects will have identical order and will be considered as equivalent, i.e., OPx ~ OPy. Note that in other studies applying partial order methodology, the term "order" is typically referred to as "rank" or "height".

Equation 1 is the basis of the Hasse diagram technique (HDT) [11,12], which is a special, statistically oriented portion of partial order theory. Thus, Hasse diagrams are visual representations of the partial orders. In the Hasse diagram, comparable objects are connected by a sequence of lines [12]. For construction of the Hasse diagram, a uniform orientation of the indicators should be secured, i.e., high indicator values correspond to "bad" objects and low values to "good" objects. In Figure 1, the Hasse diagram based on the data given in Table 1 is shown. Thus, in the present case, the higher an OP is placed in the Hasse diagram, the more hazardous and environmentally and human health damaging is it based on a simultaneous consideration of the PBT characteristics of the single OPs.

A Hasse diagram is characterized by it structure that comprises levels, chains, and anti-chains (Figure 1), where

- levels are the horizontal arrangement of objects within a Hasse diagram. Thus, the level structure gives a first approximation to a weak order of the objects from "good" (bottom) to "bad" (top).
- chains are a subset of X, such as $X' \subseteq X$, where all objects in X′ fulfill Equation 1. A chain has a length, which is $|X'|-1$. For objects within a chain (e.g., from the bottom to the top of the chain), all indicators are simultaneously non-decreasing.
- Anti-chain: A subset $X' \subseteq X$, where no object fulfills Equation 1; for example: if all objects in X′ are mutually incomparable, it is called an anti-chain. Thus, for any two objects within an anti-chain there is a conflict in indicator values.

The basic calculations were carried out applying the mHDCl2_8 module of the PyHasse software package.

2.2.3. Average orders

The level structure may be used as a first approximation to ordering. However, this will give rise to many tied orders, as all objects in a given level automatically will be assigned identical orders. Typically, it is desirable that the degree of tiedness should be as low as possible, i.e., ultimately looking for a linear ordering of the single objects. However, it is not immediately obtainable when incomparable objects are included.

Partial order methodology provides a weak order (tied orders are not excluded) by calculating the average order of the single objects [13]. In cases where only a relatively low number of objects are studied (typically < 25), an exact method for calculating average orders is available based on lattice theory [14-16]. However, for larger systems, approximations have been presented [17]. The calculations will assign an average order to the single objects. With the indicator orientation chosen as in the present study, the average orders will assigned from 1 (bottom) up to maximum n (top); n being the number of object (here 12). In the present study the average orders are calculated based on the exact method [14-16], as the study deals with only 12 objects.

The calculations to retrieve the average orders were carried out applying the avrank5 module of the PyHasse software package.

2.2.4. Perturbation of order characteristics

Each object in the Hasse diagram are characterized by what is called order characteristics, which are four values that state the number of equivalences, predecessors (number of upwards comparable objects), successors (number of downwards comparable objects), and incomparable elements, respectively [12].

Changes in one or more of the indicators may lead to changes in the order characteristics for the object under investigation. The module pooc6_3 of the PyHasse software package has been applied to visualize such changes, using MEC as an illustrative example.

2.2.5. PyHasse software

Partial order analyses were carried out applying the PyHasse software [18]. PyHasse was programmed using the interpreter language Python (version 2.6). The software package is continuously under further development and today contains more than 100 modules and is available upon request from the developer, Dr. R. Bruggemann (brg_home@web.de). The present study will apply only a few central modules such as mHDLc2_8, avrank5, pooc6_3. A simplified version of PyHasse operating on a web-based browser is under continuous development. Some tools are now available for use (www.pyhasse.org).

3. Results and Discussion

Based on the data presented in Table 1, the Hasse diagram for the 12 pesticides could be constructed (Figure 1). The diagram displays the mutual relation between the OPs as previously discussed by Sailaukhanuly *et al* [1].

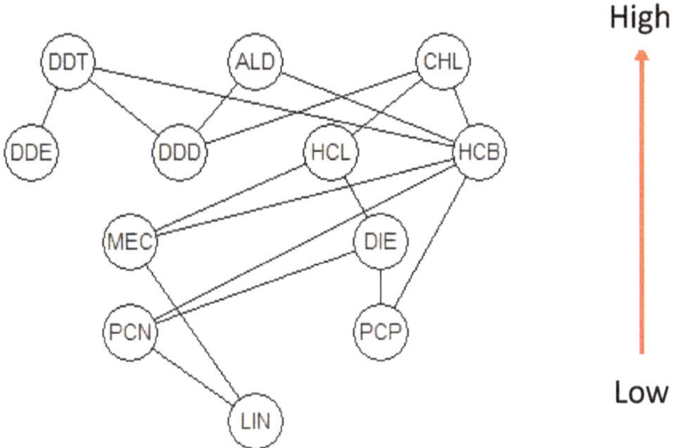

Figure 1. Hasse diagram illustrating the partial ordering of the 12 OPs given in Table 1.

From Figure 1 it can be seen that MEC is apparently incomparable to 5 OPs (DDE, DDD, DIE, PCN, PCP) and has 1 successor (LIN) and 5 predecessors (CHL, HCL, HCB, DDT, ALD). No OP was found equivalent to MEC.

In order to visualize the effect of an apparent change in the value of a specific indicator, it appears obvious to initially look at the possible changes in the order characteristics. In Figure 2, a graphic displays the summary of a perturbation of the BioA indicator by +1 (Figure 2A), as well as the resulting Hasse diagram based on the perturbed value of BioA (MEC) (Figure 2B).

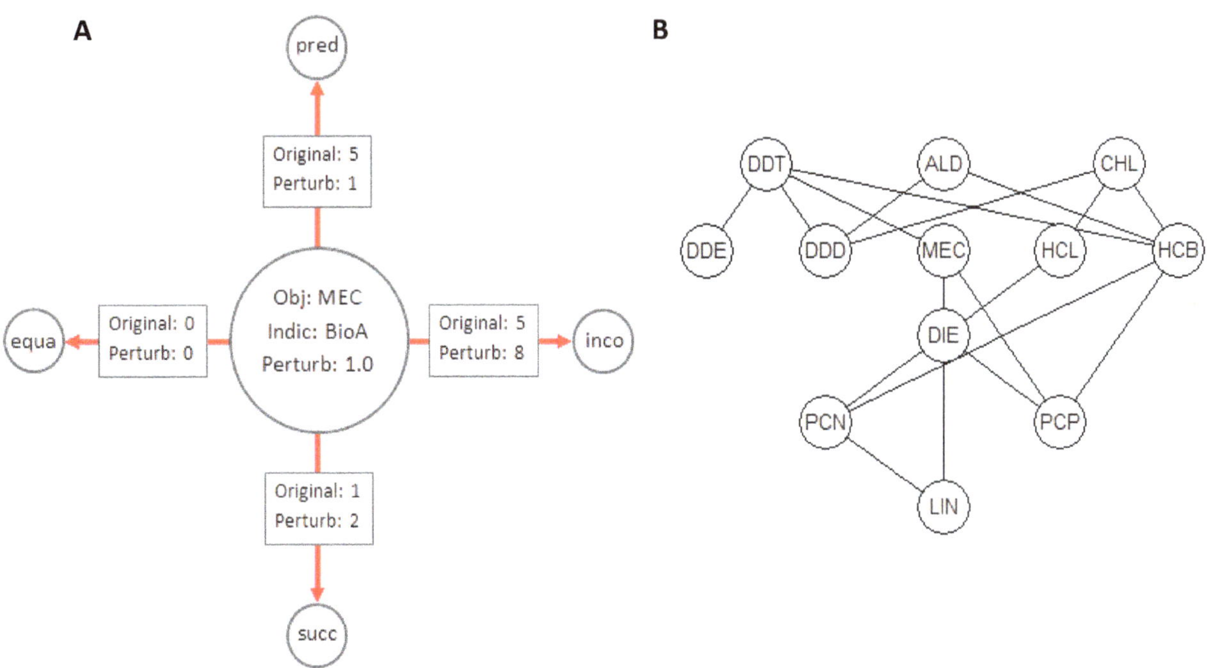

Figure 2. A: Graphics illustrating the effect of perturbation of the BioA indicator by +1 on the order characteristics of methoxychlor (MEC) (equa, pred, succ and inco denotes equalities, predecessors, successors and incomparabilities, respectively). B: The resulting Hasse diagram after perturbation.

It is immediately seen by comparing the original Hasse diagram (Figure 1) and that following the perturbation (Figure 2B), MEC is now placed higher in the diagram and is now comparable (and worse) than DIE. This is directly reflected in the number of successors that after perturbation is 2 vs 1 in the original (Figure 2A). At the same time, it is noted that the number of predecessors is reduced from 5 to 1 and the number of pesticides now incomparable to MEC is increased from 5 to 8 (Figure 2A). It should be remembered that in this context, the sum of equa, pred, succ and inco (Figure 2A) is a constant equal to the number of compounds −1. Thus, any change in one aspect must imply a reduction in another aspect. In the present example (Figure 2A), pred is heavily reduced by perturbation (from 5 to 1). Hence, this value must be compensated by the other aspects, whereby the main compensation (and the most probable) is in inco.

However, despite the significant changes in the order characteristics, a further step is needed to disclose possible synergistic and/or antagonistic effects as a function of variations in the indicator value. For this purpose, it appears logical to apply the concept of average order (alternative notation average rank or height).

To elucidate the influence, synergistic or antagonistic of changes in the apparent indicator values three sets of calculations were performed, one for each of the three indicators. Hence, with the original value of BDP3, logBCF and LC50fish as starting points, the values were varied in both positive and negative directions. All variations were, as mentioned above, carried out using MEC as an illustrative example.

3.1. Persistence

To study the effect on the ordering, with special focus on MEC, the BDP3 was varied from 0.5126 through the original value (1.5126) to 3.0126 in increments of 0.1 to disclose possible synergistic and antagonistic effects. In Figure 3, the average order of MEC as function of BDP3 (A) and the actual perturbation (B) is depicted.

It is immediately noted that decreasing the apparent BDP3 value, such as by increasing the apparent persistency of "MEC", caused a significant increase in the average order. The actual increase being from 4.3 to 7.3 and is ordered as the 4th worst OP after CHL (11.3), DDT (10.4), and ALD (10.0) (Figure 4A). These three (CHL, DDT, and ALD) still occupied the top level in the Hasse diagram (not shown), as was also the case in the unperturbed case (Figure 1).

It is further noted that a decrease in persistence only has an effect as for perturbed BDP3 values < 1.8126. For larger values of BDP3, the average order is constantly found to be 2.2, which is consistent with the fact that MEC for those values appears as a minimal element, as it has no successors (Figure 4B). In the unperturbed system, the order of MEC was found to be 4.3.

3.2. Bioaccumulation

The effect on the ordering of MEC as a function of changes in the apparent bioaccumulating of the latter was studied varying the BioA (log BCF) indicator from 0.019 through the original value (3.019) to 6.019 in increments of 0.1. The average order of MEC as function of BioA (A) and the actual perturbation (B) is depicted in Figure 5.

By perturbing the MEC value by −0.7, a drop in the average MEC order from the original (4.3) to 3.5 was caused. Further lowering the BioA did not lead to a further reduction in the average order.

It should be noted that MEC for perturbation ≤ −0.7 appears as a minimal element (Hasse diagram not shown).

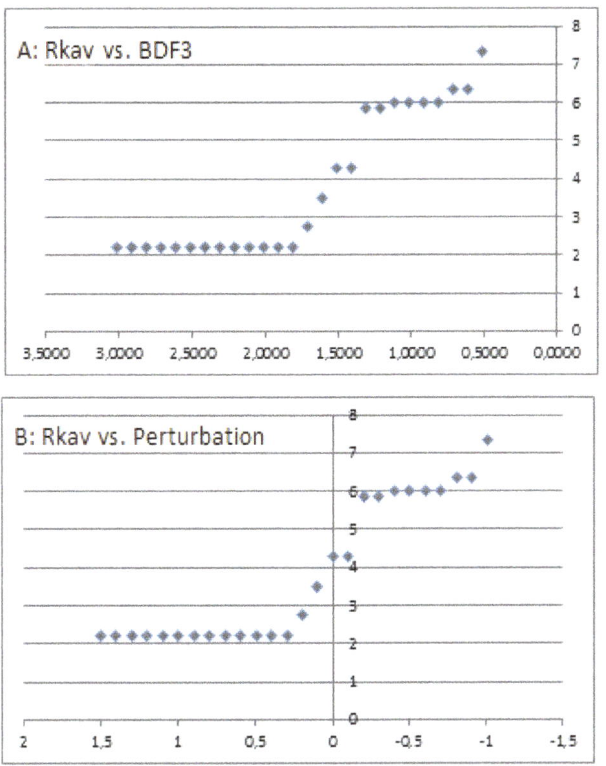

Figure 3. Average order (Rkav) vs BDP3 (A) and the actual perturbation (B). Note the reversed axis for the BDP3 and the perturbation reflecting that the lower the BDP3 the higher the persistence.

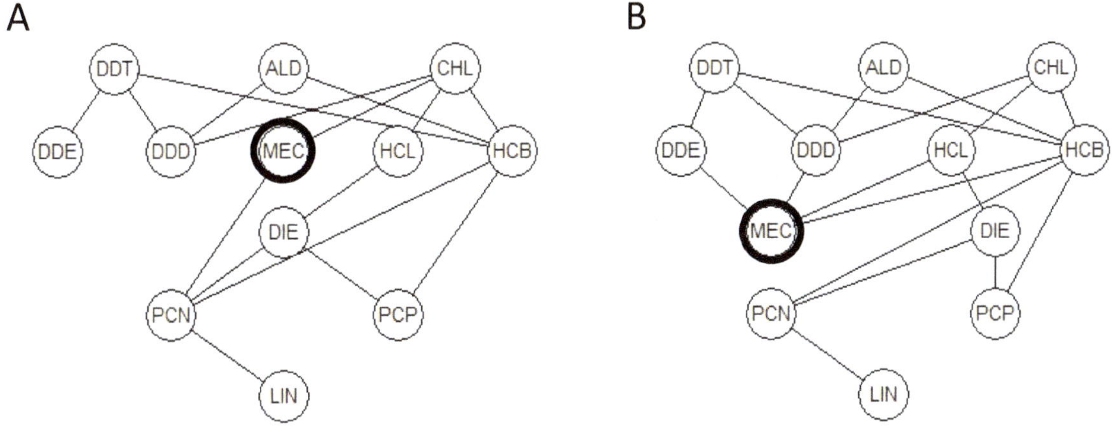

Figure 4. Hasse diagrams displaying the location of MEC following a perturbation of the BDP3 by −1 (A) and +0.3 (B).

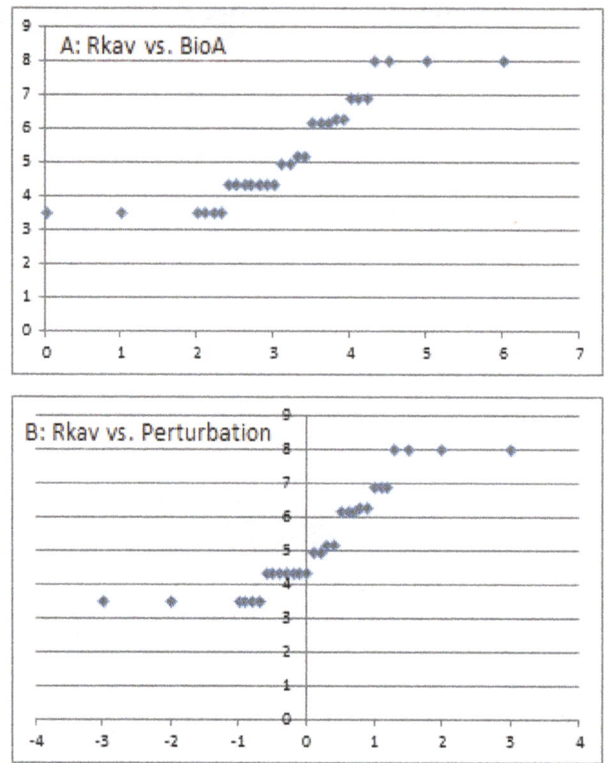

Figure 5. Average order (Rkav) vs BioA (A) and the actual perturbation (B).

On the other hand, by increasing the apparent bioaccumulation potential through the increase of the BioA indicator again lead to an increase in the average MEC order. Thus, the increase of BioA from 1.3 to 4.319 resulted in an increase in the average order to 7.9. Further increase by additional increases in the BioA was not observed. Despite the fact that MEC for BioA $\geq$ 4.319 was now a maximal element, it was still on an average basis ordered lower that DDT (10.3), CHL (11.0), and ALD (10.0).

3.3. Toxicity

The effect of a possible variation in the apparent toxicity of MEC as expressed through the LC50fish value was elucidated through a series of calculations where the LC50fish values were varied from 0.008 through the original value (0.158) to 0.308. In Figure 6, the average MEC orders as function of LC50fish (A) and the actual perturbation (B) are displayed.

It is immediately noted that decreasing the toxicity (such as by increasing the LC50fish value) did not cause any decrease of the average order of MEC, which is constantly found to be 4.3. In contrast, some increase in the average order is noted by increased toxicity as shown through decreased LC50fish. Thus, for LC50fish in the range 0.068–0.028, the average order of MEC was found to 5.6 and a further decrease in LC50fish to 0.008 caused a further increase in the average order to 7.3. In the latter case MEC is found as a maximal element. However, still fell below DDT (10.3), CHL (11.0), and ALD (10.0).

3.4. Relative indicator importance

In a previous paper [1], the relative importance of the three indicators was elucidated and it was unambiguously demonstrated that Pers >> BioA > Tox. This is in agreement with the REACH concept which states that the persistence and the bioaccumulation outweigh the toxicity if the two former indicators are high enough, i.e., for the so-called vPvB substances, where the toxicity is not brought into play [9]. However, it should be noted that this finding is an empirical finding, based on one concrete set of data. It should also be noted that the PBT criteria serves a special purpose within REACH [19].

The above results are in agreement with the previous statements. Thus, through scrutinization of Figures 3, 4, and 6, it is noted that the major change in the apparent MEC average order was seen upon variation in the BDP3, followed by BioA; whereas, even significant changes in the LC50fish indicator lead only to relatively smaller variations in the MEC average order.

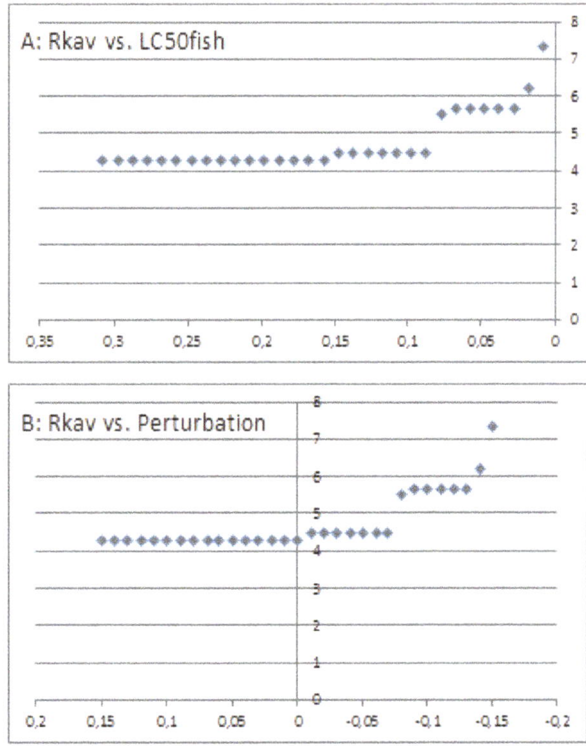

Figure 6. Average order (Rkav) vs LC50fish (A) and the actual perturbation (B). Note the reversed axis for the LC50fish and the perturbation reflecting that the lower the LC50fish the higher the toxicity.

4. Conclusions

The possible synergistic and antagonistic effects on the relative hazard of chemicals have been elucidated applying simple partial order methodology. As an illustrative example, the relative hazard of methoxychlor (MEC) in a group of 12 obsolete chloro-containing pesticides was used.

The 12 pesticides were partially ordered by applying their persistence, bioaccumulation, and toxicity as indicators. The average order of MEC was estimated as function of a systematic variation in the three indicators. The ordering of the pesticides by simultaneously including these 3 indicators is believed to constitute an ordering according to the environmental hazard [19]. The present study thus demonstrates the advantageous use of partial ordering, which is an analytical approach that allows a simultaneous inclusion of all applied indicators. Partial ordering does not require any assumption on linearity or distribution of the indicators and such available data can be applied immediately. Hence, the information contained in the single indicator is maintained and not being hidden by a possible aggregation process, which could infer compensation effects [20].

It was found that increasing either the persistence, the bioaccumulation, or the toxicity caused a relative increase in the average order of MEC compared to the other 11 pesticides, resulting in an increased hazard. However, a decrease in the indicator values lead to a decreased relative hazard, apart from in the toxicity case, where a decrease in toxicity did not apparently cause any change in the relative average order of MEC.

It was further shown that the observed results are consistent with previous studies which demonstrated that the persistence and bioaccumulation indicators are the two dominating figures, but toxicity plays a somewhat less dominant role. This is further in agreement with the REACH regulation, where a distinction is made between PBT and vPvB compounds, which is to say that if persistence and bioaccumulation are both sufficiently high, the toxicity indicators are not taken into account.

In the present study, only the variation in one indicator at a time for one single compound, MEC, has been dealt with. Hence, a larger systematic study where all possible variations for single or multiple compounds and single or multiple indicators would be straightforward, but outside the scope of the present paper. However, the present study unambiguously demonstrates how partial order methodology can be applied as an advantageous tool in elucidating and quantifying synergistic and antagonistic effects which cause increased or decreased hazard of chemicals due to the possible presence of 'third party' agents. Such analyses thus contribute to the current discussion on the possible unwanted effects caused by factors such as nanomaterials and microplastics.

The partial order analytical tool presented here may serve as an initial analysis prior to elaborate, time consuming, a priori, and rather costly experimental studies.

Conflicts of interest

Author declares no conflicts of interest in this paper.

References

1. Sailaukhanuly Y, Zhakupbekova A, Amutova F, et al. (2013) On the Ranking of Chemicals based on their PBT Characteristics: Comparison of Different Ranking Methodologies Using Selected POPs as an Illustrative Example. *Chemosphere* 90: 112-117.
2. Steinberg ChEW, Haitzer M, Geyer HJ (2001) Biokonzentrations studien in Abhängigkeit von Huminstoffen, In: *Streß in limnischen Ökosystemen. Neu Ansätze in der ökotoxikologischen Bewartung von Binnengewässern, Steinberg*, Ch.E.W., Brüggemenn, R., Kümmerer-Liess, K., Pflugmacher, St and Zauge, G.-P. (eds), Parey Buchverlag, Berlin, 2001, section 2.2, pp. 16-25.
3. Beat the microbead (2014) Microplastics: scientific evidence, http://beatthemicrobead.org/en/science (accessed Nov, 2014).

4. Syberg K (2009) Mixture toxicity and European chemical regulation: a interdisciplinary study, PhD thesis, Roskilde University, 88pp.
5. Stockholm Convention (2008) http://chm.pops.int/Home/tabid/2121/language/en-GB/Default.aspx (accessed Nov. 2014).
6. Mallouk TE, Yang P (2009) Chemistry at the Nano-Bio Interface. *J Am Chem Soc* 131: 7937-7939.
7. EPA (2013) Estimation Program Interface (EPI) Suite, ver. 4.11 http://www.epa.gov/oppt/exposure/pubs/episuite.htm (accessed Nov. 2014).
8. Walker JD, Carlsen L (2002) QSARs for identifying and prioritizing substances with persistence and bioaccumulation potential, SAR QSAR. *Environ Res* 13: 713-726.
9. Koch V (2008) PBT Assessment & Category Approach, http://reach.setac.eu/embed/presentations_reach/18_KOCH_PBT_Assessment.pdf (accessed Nov. 2014).
10. Bruggemann R, Carlsen L (2012) Multi-criteria decision analyses. Viewing MCDA in terms of both process and aggregation methods: Some thoughts, motivated by the paper of Huang, Keisler, and Linkov. *Sci Tot Environ* 425: 293-295.
11. Bruggemann R, Carlsen L (2006) *Partial Order in Environmental Sciences and Chemistry*. Springer, Berlin.
12. Bruggemann R, Patil GP (2011) *Ranking and prioritization for multi-indicator systems—Introduction to partial order applications.* Springer, New York.
13. Bruggemann R, Annoni P (2014) Average heights in partially ordered sets. *MATCH Commun Math Comput Chem* 71: 117-142.
14. De Loof K, De Meyer H, De Baets B (2006) Exploiting the Lattice of Ideals Representation of a Poset. *Fundamenta Informaticae* 71: 309-321.
15. Morton J, Pachter L, Shiu A, et al. (2009) Convex Rank Tests and Semigraphoids. *SIAM J Discrete Math* 23: 1117-1134.
16. Wienand O (2006) lcell, http://bio.math.berkeley.edu/ranktests/lcell/ (accessed Nov 2014).
17. Bruggemann R, Carlsen L (2011) An Improved Estimation of Averaged Ranks of Partially Orders. *MATCH Commun Math Comput Chem* 65: 383-414.
18. Bruggemann R, Carlsen L, Voigt K, et al. (2014) PyHasse Software for Partial Order Analysis. In R. Bruggemann, L. Carlsen, J. Wittmann (Eds.), *Multi-Indicator Systems and Modelling in Partial Order* (pp. 389–423). Springer, New York.
19. EC (2006) Regulation (EC) No 1907/2006 of the European Parliament and of the Council of 18 December 2006 concerning the Registration, Evaluation, Authorisation and Restriction of Chemicals (REACH), establishing a European Chemicals Agency, amending Directive 1999/45/EC and repealing Council Regulation (EEC) No 793/93 and Commission Regulation (EC) No 1488/94 as well as Council Directive 76/769/EEC and Commission Directives 91/155/EEC, 93/67/EEC, 93/105/EC and 2000/21/EC; Article 57d,e (http://eur-lex.europa.eu/legal-content/EN/TXT/?qid=1415770830954&uri=CELEX:32006R1907); Accessed Nov. 2014)
20. Munda G (2008) Social Multi-Criteria Evaluation for a Sustainable Economy. Springer-Verlag, Berlin.

7

Leading edge technologies in wastewater treatment

Rafael Gonzalez-Olmos

IQS School of Engineering, Universitat Ramon Llull, Via Augusta, 390, 08017 Barcelona, Spain

In this special edition of *AIMS Environmental Science*, we highlight different aspects of some advanced (leading edge) technologies in the field of wastewater treatment. Water is one of the most valuable resources of our planet and in the following years the concepts of water treatment and water reuse will be more relevant than even today. The technological research trends in this field are closely related to the new challenges of water science. As an example, currently new technologies are required for the efficient treatment of micropollutants such as pharmaceuticals or for reaching the quality standards for water reuse [1]. Other challenges are related with the need of technologies that can help us to extract the different resources enclosed into the wastewater such as nutrients [2] or to reduce greenhouse gases emissions [3] from wastewater treatment facilities. In this issue, a series of different articles try to cover, on one hand, specific advanced areas which are well established such as membrane bioreactors (MBR) or moved bed bioreactor (MBBR). On the other hand, other emerging technologies which are not widely used in full-scale facilities are analyzed as the application of heterogeneous Fenton-like processes or the use of the biogeochemical cycle of iron to recover resources from wastewater.

Wastewater treatment has been from its birth a field in continuous evolution. Since 100 years ago, when Ardern and Lockett discovered the activated sludge process [4], several advances have been done with the objective of reaching the requirements always demanded by the authorities and society. Despite of these advances and that activated sludge is one of the most important biotechnological processes used around in the world, still there remain significant gaps related to some aspects such as the flocs formation. Aspects related to how the floc development is initiated or which are the relative contributions of cell replication and cell recruitment to floc growth are poorly described. A promising approach to addressing these knowledge gaps is studied in this special issue [5].

MBR is an advanced technology which is currently highly implemented in different urban and industrial wastewater treatment facilities [6]. MBR technology is a step forward in the classical activated sludge treatment. In the activated sludge treatment the solid-liquid separation unit is a settler where the suspended solids are separated from the effluent by sedimentation. In MBR, membrane separation processes replace the settler units. With this replacement several goals are achieved. The first one is that the operational surface needed to carry out the solid-liquid separation is significantly reduced. The second goal is that the quality of the effluents is also increased. The main drawback related with MBR are the necessity of using high amounts of energy and also the operational problems derived from the operation and maintenance of the membranes [7]. In this special issue, all these aspects are faced in a full-scale MBR water reuse facility [8].

In the last 20 years, MBBR has been established as a simple-yet-robust, flexible and compact technology for wastewater treatment [9]. It has shown great potential in pollution load reduction and has definite edge over the surface aeration system [10] and its application in wastewater treatment has increased over the past decade. MBBRs have become an interesting alternative for wastewater treatment as it is a reliable and compact system due to development in the design and operation which has resulted in decreased footprints, significantly lower suspended solid production, consistent production of high quality and reusable water and minimal waste disposal. It is one of the advanced aerobic wastewater treatment processes having advantages of both attached and suspended growth systems. In MBBR the biomass grows as a biofilm on small plastic carriers that move freely into the wastewater [11]. A screen is provided at the outfall end of the reactor to keep media from passing out of the reactor. Contrary to the activated sludge reactor, it does not need any sludge recycle. Since no sludge recirculation takes place, only the surplus biomass has to be separated, which is a considerable advantage over the activated sludge process [12]. Moreover, nitrification and de-nitrification can also be successfully achieved in biofilm-based processes since nitrifiers, which are slow growing micro-organisms, are retained by the biofilm [13]. Several processes with different kinds of floating carriers have been developed [14], with porous materials such as polyurethane and with nonporous material such as polyethylene. In this issue it will be presented a study related with the upgrading of a full scale wastewater treatment plant to MBBR [15].

The removal of organic matter from contaminated effluents usually is removed biologically in activated sludge systems, or in more advanced systems such as MBR or MBBR. The removal in all those treatments is carried out biologically, so only that fraction of the organic matter which is really biodegradable can be removed. In the case of industrial wastewaters, the concentration of refractory or non-biodegradable organic compounds used to be high. The presence of these kind of compounds in wastewaters, which can be carcinogenic or/and mutagenic, with high toxicological potentials, is a global issue because of the increasing demand for public health awareness and environmental quality. So the development of quick and simple methods for their removal are still required. Advanced Oxidation Processes (AOP's) have emerged as interesting alternative for the destruction of organic pollutants in industrial wastewater [16]. These processes involve the generation of non-selective and highly reactive hydroxyl radicals, which are one of the most powerful oxidation agents. The reaction of hydrogen peroxide with ferrous salts (Fenton's reagent) or other low-valence transition metals (Fenton-like reactions) is a well-known source of hydroxyl radicals at room temperature [17]. In this issue, it will be studied how to overcome the main drawbacks of the Fenton process by using heterogeneous catalysts [18].

New technologies based on the study of the biogeochemical cycles can have a big impact in the future of the wastewater treatment plants. One example based in the biogeochemical cycle of iron is the BioIronTech process which is covered by an article in this special issue [19]. BioIronTech process have shown to remove/recover more than 90% of phosphorous from reject water thus replacing the conventional process of phosphate precipitation by ferric/ferrous salts, which are

20-100 times more expensive than iron ore, which is used in BioIronTech process. BioIronTech process can remarkably improve the aerobic and anaerobic treatments of municipal and industrial wastewaters, especially anaerobic digestion of lipid- and sulphate-containing food-processing wastewater. It can also remove the recalcitrant compounds from industrial wastewater, enhance sustainability and quality of water resources, and recover ammonium and phosphate from municipal and food-processing wastes.

Overall this review series provides a snapshot of the power of leading edge technologies and concepts which can be relevant in the future wastewater treatment facilities. Obviously, the field of leading edge technologies related to water is really wide and is difficult to cover all the areas which are in development in different research centers around the world. Some of the areas not covered in this review and that deserve to be mentioned are some emerging areas related with the resource recovery from wastewater such as bioplastics [20], nutrients in form of different fertilizers such as struvite [21] or the energy contained into the organic matter using biolectrochemical systems [22]. Topics related with micropollutants removal [23] and water reuse [24] that will be highly relevant in future were also not covered in this special issue. Undoubtedly, these new areas, as well as further research into the topics covered in this review series, will provide further advances in the development of water treatments

Acknowledgement

R. Gonzalez-Olmos thanks to "Obra Social La Caixa" for receiving funding to carry out this research through the University Ramon Llull Tractor project "Electromembrane" (2014-URL-Trac-002). Obra Social "la Caixa"

Conflict of Interest

Author declares no conflicts of interest in this paper.

References

1. Drewes JE, Reinhard M, Fox P (2003) Comparing microfiltration-reverse osmosis and soil-aquifer treatment for indirect potable reuse of water. *Water Res* 37: 3612-3621.
2. Guest JS, Skerlos SJ, Barnard JL, et al. (2009) A new planning and design paradigm to achieve sustainable resource recovery from wastewater, *Environ Sci Technol* 43: 6126-6130.
3. Kampschreur MJ, Temmink H, Kleerebezem R, et al. (2009) Nitrous oxide emission during wastewater treatment. *Water Res* 43: 4093-4103.
4. Ardern E, Lockett WT (1914) Experiments on the oxidation of sewage without the aid of filters. *J Soc Chem Ind* 33: 523-539.
5. Kretzschmar AL, Manefield M (2015) The role of lipids in activated sludge floc formation. *AIMS Environ Sci* 2: 122-133.
6. Melin T, Jefferson B, Bixio D (2006) Membrane bioreactor technology for wastewater treatment and reuse. *Desalination* 187: 271-282.
7. Verrecht B, Judd SJ, Guglielmi G (2008) An aeration energy model for an immersed membrane

bioreactor. *Water Res* 42: 4716-4770.

8. Wen J, Liu Y, Tu Y, et al. (2015) Energy and chemical efficient nitrogen removal at a full-scale MBR water reuse facility. *AIMS Environ Sci* 2: 42-55.
9. Jenkins AM, Sanders D (2012) Introduction to fixed-film bioreactors for decentralized wastewater treatment. *Contech, Engineered Solutions*. http://www.conteches.com/knowledge-center/pdh-article-series/introduction-to-fixed-film-bio-reactors.aspx.
10. Das A, Naga RN (2011) Activated sludge process with MBBR technology at ETP. *Ippta J* 23: 135-137.
11. Qiqi Y, Qiang H, Ibrahim HT (2012) Review on moving bed biofilm processes. *Pak J Nutr* 11: 706-713.
12. Ødegaard H (2006) Innovations in wastewater treatment: the moving bed biofilm process. *Water Sci Technol* 53*:* 17-33.
13. Aygun A, Nas B, Berktay A (2008) Influence of high organic loading rates on COD removal and sludge production in moving bed biofilm reactor. *Environ Eng Sci* 25: 1311-1316
14. Pastorelli G, Andreottola G, Canziani R, et al. (2000) Pilot plant experiments with moving bed biofilm reactors. *Water Sci Technol* 36: 43-50
15. Falletti L, Conte L, Maestri A (2014) Upgrading of a wastewater treatment plant with a hybrid moving bed biofilm reactor (MBBR*). AIMS Environ Sci* 2: 45-52.
16. Neyens E, Baeyens J (2003) A review of classic Fenton's peroxidation as an advanced oxidation technique. *J Hazard Mater* 98: 33-50.
17. Georgi, A., Gonzalez-Olmos, R., Köhler, R., Kopinke, F.-D. Fe-zeolites as catalysts for wet peroxide oxidation of organic groundwater contaminants: Mechanistic studies and applicability tests. *Sep. Sci. Technol.* 45:1579-1586.
18. Pariente MI, Martínez F, Botas JÁ, et al. (2015) Extrusion of Fe2O3/SBA-15 mesoporous material for application as heterogeneous Fenton-like catalyst. *AIMS Environ Sci* 2: 154-168.
19. Ivanov V, Stabnikov V, Guo CH, et al. (2014) Wastewater engineering applications of BioIronTech process based on the biogeochemical cycle of iron bioreduction and (bio)oxidation. *AIMS Environ Sci* 1: 53-66.
20. Johnson K, Kleerebezem R, van Loosdrecht MCM (2010) Influence of the C/N ratio on the performance of polyhydroxybutyrate (PHB) producing sequencing batch reactors at short SRTs. *Water Res* 44: 2141-2152.
21. Liu Y, Kumar S, Kwag JH, et al. (2013) Magnesium ammonium phosphate formation, recovery and its application as valuable resources: A review. *J Chem Technol Biotechnol* 88: 181-189.
22. Puig S, Serra M, Coma M, et al. (2011) Simultaneous domestic wastewater treatment and renewable energy production using microbial fuel cells (MFCs). *Water Sci Technol* 64: 904-909.
23. Collado N, Rodriguez-Mozaz S, Gros M, et al. (2014) Pharmaceuticals occurrence in a WWTP with significant industrial contribution and its input into the river system. *Environ Poll* 185: 202-212.
24. Shaffer DL, Arias Chavez LH, Ben-Sasson M, et al. (2013) Desalination and reuse of high-salinity shale gas produced water: Drivers, technologies, and future directions. *Environ Sci Technol* 47: 9569-9583.

8

Linking state-and-transition simulation and timber supply models for forest biomass production scenarios

Jennifer K. Costanza [1,2], Robert C. Abt [2], Alexa J. McKerrow [3] and Jaime A. Collazo [4]

[1] North Carolina Cooperative Fish and Wildlife Research Unit, Department of Applied Ecology, North Carolina State University, Raleigh, NC 27695, USA

[2] Department of Forestry and Environmental Resources, North Carolina State University, Raleigh, NC 27695, USA

[3] Core Science Analytics and Synthesis, U.S. Geological Survey, Raleigh, NC 27695, USA

[4] U.S. Geological Survey, North Carolina Cooperative Fish and Wildlife Research Unit, Department of Applied Ecology, North Carolina State University, Raleigh, NC 27695, USA

Abstract: We linked state-and-transition simulation models (STSMs) with an economics-based timber supply model to examine landscape dynamics in North Carolina through 2050 for three scenarios of forest biomass production. Forest biomass could be an important source of renewable energy in the future, but there is currently much uncertainty about how biomass production would impact landscapes. In the southeastern US, if forests become important sources of biomass for bioenergy, we expect increased land-use change and forest management. STSMs are ideal for simulating these landscape changes, but the amounts of change will depend on drivers such as timber prices and demand for forest land, which are best captured with forest economic models. We first developed state-and-transition model pathways in the ST-Sim software platform for 49 vegetation and land-use types that incorporated each expected type of landscape change. Next, for the three biomass production scenarios, the SubRegional Timber Supply Model (SRTS) was used to determine the annual areas of thinning and harvest in five broad forest types, as well as annual areas converted among those forest types, agricultural, and urban lands. The SRTS output was used to define area targets for STSMs in ST-Sim under two scenarios of biomass production and one baseline, business-as-usual scenario. We show that ST-Sim output matched SRTS targets in most cases. Landscape dynamics results indicate that, compared with the baseline scenario, forest biomass production leads to more forest and, specifically, more intensively managed forest on the landscape by 2050. Thus, the STSMs, informed by forest economics models, provide important information about potential landscape effects of bioenergy production.

Keywords: bioenergy; forest dynamics; forest residues; land-use change; landscape dynamics; longleaf pine ecosystem; Southeastern U.S.; ST-Sim; SubRegional Timber Supply (SRTS) model; timber harvest

1. Introduction

Recently, there has been much discussion about the potential for bioenergy production from forest-derived biomass. For example, forests could provide important sources of cellulosic feedstocks for second-generation liquid biofuels [1]. In addition, wood pellets from forest biomass may be a critical part of the global renewable electricity portfolio in the near future [2]. Whether for liquid biofuels or wood pellets, forest biomass may lower some ecological impacts compared with crop-based biofuel feedstocks, which are usually grown in monocultures [3]. However, compared with energy sources such as coal and petroleum, cellulosic biofuels and electricity from biomass require more land per unit of energy [4]. Therefore, widespread forest biomass production for bioenergy has the potential to cause substantial land-use change, and thus impact landscapes, habitats, and ecosystems [5-7].

In the southeastern U.S., biomass from forests could be an important source of bioenergy [8-10]. Forest biomass production in the region would likely lead to increased conversion of naturally-regenerating forests and agricultural lands to intensively managed forests, elevated thinning and harvest rates in all forests, and increased use of forest residues such as small trees and branches [9]. While several studies have investigated ecological impacts of these types of changes at local scales, the landscape impacts of large-scale biomass production has been identified as a critical research need in the Southeast and beyond [3,10,11]. Investigating landscape impacts and tradeoffs using plausible future scenarios of biomass production over time is a particularly important research need [12]. In the Southeast, future scenarios that incorporate forest economics will be especially critical because factors like demand for forest products and forest land are key drivers of land change there [13,14]. Here, we model landscape dynamics of potential forest biomass production scenarios in North Carolina using state-and-transition simulation models (STSMs), along with an economics-based timber supply model.

Several features of STSMs make them ideal for simulating land conversion and elevated management that are likely under alternative biomass production scenarios. In STSMs, vegetation stages, land-use types, and transitions among them, whether due to ecological or anthropogenic drivers, are represented by discrete model components much like Markov models or conceptual state-and-transition models [15,16]. But, additional features in STSMs also facilitate the simulation of transitions such as land-use change or management regimes that can change over time [15]. Furthermore, unlike other simulation models built exclusively for forest vegetation types, the flexibility of STSMs allows representation of any vegetation or anthropogenic land-use type. For these reasons, STSMs have become increasingly popular tools for investigating scenarios of landscape and forest dynamics [17-20] and make them well-suited for our study.

Our objective in this study was to link STSMs with an economics-based timber supply model to examine landscape dynamics in North Carolina (NC) through 2050 under two scenarios of forest biomass production along with a baseline, or business-as-usual scenario. The SubRegional Timber Supply (SRTS) model [21,22], was used to incorporate the important economic drivers of forest land change and management into STSMs. Specifically, our objectives were to: (1) simulate potential future landscape dynamics in NC using STSMs based on output from a timber supply model, and (2) examine how alternative simulated scenarios of forest biomass production are likely to affect landscape dynamics in NC. To address our first objective, we used SRTS to generate annual target areas for land-use conversion and forest management transitions, then input these targets into an STSM and examined how closely target areas from SRTS matched areas simulated by the STSM. To address our second objective, we compared the three scenarios in terms of areas of vegetation and land-use types in NC through time, and annual areas experiencing transitions such as forest disturbance and management. The result is a new application of STSMs for understanding potential landscape dynamics due to bioenergy production.

2. Materials and Methods

We linked timber supply and state-and-transition simulation models (STSMs) to examine the effects of forest biomass production for bioenergy in NC using the following approach. First, we used supply chain analysis and existing literature to determine how much demand for forest biomass might be expected in the future. We developed two scenarios of forest biomass production to fit that demand, along with a third, baseline, or business-as-usual scenario. Second, we developed a series of state-and-transition model pathways in the STSM software platform ST-Sim to represent ecological dynamics, forest management, and land-use change within and among vegetation and land-use types across the state. Third, we used recent land cover and other spatial data to determine the composition of vegetation (forest and non-forest) and land-use types across North Carolina in the year 2010. Fourth, we used the SubRegional Timber Supply (SRTS) model to project areas of forest that are thinned or harvested under each scenario, along with changes in the area of major forest types, urban lands, and agricultural lands across four regions of the state. Fifth, we input the SRTS projections, and initial conditions into ST-Sim to project changes in land use and vegetation from 2011 through 2050.

For modeling, we divided NC into four regions that are analogous to those used by the USDA Forest Service's Forest Inventory and Analysis (FIA) program (Figure 1): groups of counties corresponding to the Northern Coastal Plain, Southern Coastal Plain, Piedmont and Mountains. The SRTS model is designed to work at regional extents, and these are regions of similar land use and vegetation, with relatively homogenous drivers of change (see below). In addition, the regional domains allowed reasonably fast simulation by the STSMs.

2.1. Study area: North Carolina

The state of NC has three distinct physiographic regions: the Coastal Plain, Piedmont, and Mountains. Major forested vegetation types across the state include longleaf pine ecosystems and oak-pine forests (Table 1). The majority (89%) of the state is privately owned, with most large public areas occurring in the Mountains and on the outer Coastal Plain [23]. Planted loblolly pine (*Pinus taeda*) plantations make up 8% of the state, and much of that area (58%) is in the Coastal Plain region. Agricultural land uses, including row crops and pastures, cover 23% of the state's area, most of which is in the Coastal Plain (56%) and Piedmont (35%). Urban areas occur across the state, making up 11% of the land use statewide. The majority (52%) of urban land is in the Piedmont region, which contains the state's largest urban areas, including Charlotte and the Raleigh-Durham area. The Piedmont region is also projected to see the highest future rate of urbanization [24]. The state contains some of the most highly-diverse ecosystems in the world, and in particular, the Coastal Plain occurs in a global biodiversity hotspot [25]. Therefore, it is important to determine how biomass production will affect landscapes in NC. To meet future energy demand in NC, forest biomass will likely be produced from the major forest types, including managed and non-managed pine and hardwood forests, which are already present on the landscape, and are used to produce other forest-based products such as timber and paper products.

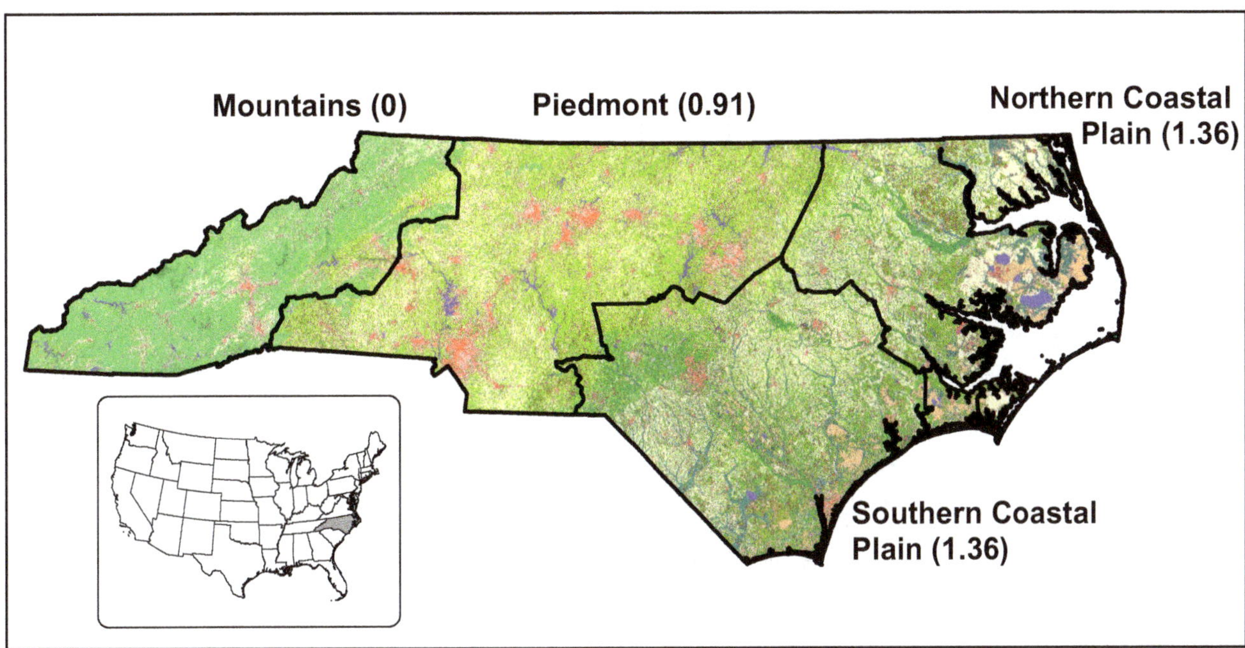

Figure 1. Map of North Carolina, showing the four simulation regions in the state along with recent mapped land cover [26]**.** Numbers in parentheses are quantities of biomass, in million green tonnes, produced annually for bioenergy starting in 2019 under both biomass scenarios. In the land cover, green colors indicate forested ecosystems, peach indicates wetlands, blues indicate water, tans indicate agriculture, and reds indicate urban lands.

Table 1. Information about the land-use and vegetation types modeled and their initial extents.

Name	Initial area (ha)	% of total extent	SRTS general forest type[a]	General type	Model source[b]
Appalachian Hemlock-Hardwood Forest	8,927	0.09%	Mixed pine hardwood	Forest	LANDFIRE
Atlantic Coastal Plain Central Fresh-Oligohaline Tidal Marsh	4,036	0.04%	NA	Wetland	LANDFIRE
Atlantic Coastal Plain Central Maritime Forest	5,277	0.05%	NA	Forest	LANDFIRE
Atlantic Coastal Plain Central Salt and Brackish Tidal Marsh	14,142	0.14%	NA	Wetland	LANDFIRE
Atlantic Coastal Plain Clay-Based Carolina Bay Forested Wetland	3,581	0.04%	NA	Forest	LANDFIRE
Atlantic Coastal Plain Dry and Dry-Mesic Oak Forest	387,324	3.87%	Upland hardwood	Forest	LANDFIRE
Atlantic Coastal Plain Embayed Region Tidal Freshwater Marsh	2,639	0.03%	NA	Wetland	LANDFIRE
Atlantic Coastal Plain Embayed Region Tidal Salt and Brackish Marsh	33,485	0.33%	NA	Wetland	LANDFIRE
Atlantic Coastal Plain Fall-line Sandhills Longleaf Pine Woodland	137,623	1.37%	Natural pine	Forest	LANDFIRE
Atlantic Coastal Plain Mesic Hardwood and Mixed Forest	38,982	0.39%	Upland hardwood	Forest	LANDFIRE
Atlantic Coastal Plain Nonriverine Swamp and Wet Hardwood Forest	123,793	1.24%	NA	Forest	LANDFIRE
Atlantic Coastal Plain Northern Basin Swamp and Wet Hardwood Forest	8,451	0.08%	Lowland hardwood	Forest	LANDFIRE
Atlantic Coastal Plain Northern Dune and Maritime Grassland	232	0.00%	NA	Wetland	LANDFIRE
Atlantic Coastal Plain Northern Fresh and Oligohaline Tidal Marsh	216	0.00%	NA	Wetland	LANDFIRE
Atlantic Coastal Plain Northern Maritime Forest	435	0.00%	NA	Forest	LANDFIRE
Atlantic Coastal Plain Northern Tidal Salt Marsh	7,165	0.07%	NA	Wetland	LANDFIRE
Atlantic Coastal Plain Northern Tidal Wooded Swamp	711	0.01%	Lowland hardwood	Forest	LANDFIRE
Atlantic Coastal Plain Peatland Pocosin	247,828	2.47%	NA	Forest	LANDFIRE
Atlantic Coastal Plain Small Blackwater River Floodplain Forest	268,943	2.68%	Lowland hardwood	Forest	LANDFIRE
Atlantic Coastal Plain Small Blackwater Stream Floodplain Forest	129,921	1.30%	Lowland hardwood	Forest	LANDFIRE
Atlantic Coastal Plain Small Brownwater Floodplain Forest	150	0.00%	Lowland hardwood	Forest	LANDFIRE
Atlantic Coastal Plain Small Brownwater River Floodplain Forest	100,975	1.01%	Lowland hardwood	Forest	LANDFIRE
Atlantic Coastal Plain Southern Dune and Maritime Grassland	3,392	0.03%	NA	Grass/shrub	LANDFIRE
Atlantic Coastal Plain Southern Tidal Wooded Swamp	4,843	0.05%	Lowland hardwood	Forest	LANDFIRE
Atlantic Coastal Plain Streamhead Seepage Swamp, Pocosin, and Baygall	2,233	0.02%	NA	Forest	LANDFIRE
Atlantic Coastal Plain Upland Longleaf Pine Woodland	676,615	6.75%	Natural pine	Forest	LANDFIRE

Central and Southern Appalachian Montane Oak Forest	40,591	0.41%	Upland hardwood	Forest	LANDFIRE
Central and Southern Appalachian Northern Hardwood Forest	15,132	0.15%	Upland hardwood	Forest	LANDFIRE
Central and Southern Appalachian Spruce-Fir Forest	199	0.00%	Natural pine	Forest	LANDFIRE
Central Atlantic Coastal Plain Wet Longleaf Pine Savanna and Flatwoods	45,846	0.46%	Natural pine	Forest	LANDFIRE
Pasture/Hay	1,484,970	14.82%	NA	Agriculture	New model
Planted Pine - Private	1,068,082	10.66%	Planted pine	Forest	New model
Row Crop	1,623,992	16.21%	NA	Agriculture	New model
South-Central Interior Large Floodplain	281	0.00%	Lowland hardwood	Forest	LANDFIRE
South-Central Interior Small Stream and Riparian	2,927	0.03%	Lowland hardwood	Forest	LANDFIRE
Southeastern Interior Longleaf Pine Woodland	88	0.00%	Natural pine	Forest	LANDFIRE
Southern and Central Appalachian Cove Forest	41,855	0.42%	Upland hardwood	Forest	LANDFIRE
Southern and Central Appalachian Oak Forest	545,218	5.44%	Upland hardwood	Forest	LANDFIRE
Southern and Central Appalachian Oak Forest - Xeric	306,289	3.06%	Upland hardwood	Forest	LANDFIRE
Southern Appalachian Grass and Shrub Bald	138	0.00%	NA	Grass/shrub	LANDFIRE
Southern Appalachian Low Mountain Pine Forest	23,445	0.23%	Natural pine	Forest	LANDFIRE
Southern Appalachian Montane Pine Forest and Woodland	694	0.01%	Natural pine	Forest	LANDFIRE
Southern Piedmont Dry Oak Pine Forest	2,065,837	20.62%	Mixed pine hardwood, natural pine[c]	Forest	LANDFIRE
Southern Piedmont Large Floodplain Forest	5,223	0.05%	Lowland hardwood	Forest	LANDFIRE
Southern Piedmont Mesic Forest	748	0.01%	Upland hardwood	Forest	LANDFIRE
Southern Piedmont Small Floodplain and Riparian Forest	42,428	0.42%	Lowland hardwood	Forest	LANDFIRE
Southern Ridge and Valley Dry Calcareous Forest - Pine	7	0.00%	Natural pine	Forest	LANDFIRE
Urban[d]	466,828	4.66%	NA	Urban	New model
Total	9,992,735	–	–	–	–

[a] Non-forested vegetation types as well as forest types that are unlikely to be managed for biomass were excluded from SRTS. See text for more details.

[b] "LANDFIRE" indicates that the modified from a model developed for the same system by LANDFIRE (http://www.landfire.gov).

[c] In the Southern Piedmont Dry Oak Pine Forest model, the state classes corresponding to pine-dominated conditions were treated as natural pine forest. All others were mixed pine.

[d] Urban land in initial conditions did not include land that was classified developed but with low intensity or open space.

2.2. Scenarios of future demand for forest biomass

North Carolina has the potential for substantial forest-based biomass demand for bioenergy. The state is the only one in the southeastern U.S. with a renewable energy portfolio standard and, until recent changes in state policy, had a biofuel target of 10% of liquid transportation fuels from locally produced biofuels [27]. Such demand could be met from forest biomass, or from agricultural crops such as switchgrass or sweet sorghum. In addition to biofuels, existing or announced wood pellet plants in NC could consume 3.45 million green tonnes (3.8 million green short tons) of forest biomass annually [28]. For this study we examined the impact of producing 3.63 million green tonnes (4.0 million green short tons) of wood-based bioenergy annually, which would meet up to 29% of the biofuels target (see Supplementary Material for biomass to ethanol conversion) or 100% of wood pellet demand.

We developed two scenarios that incorporated annual production of 3.63 million green tonnes of biomass for bioenergy but varied in the degree to which forest residues would be removed and compared those to a business-as-usual scenario in which no biomass was produced for bioenergy. Forest residues include small trees, tree tops, rough wood, rotten logs and snags that cannot be used in a conventional lumber mill or other processing facility. Because forest residues play important ecological roles, such as providing habitat for wildlife, promoting biodiversity, and as sources of soil nutrients, the degree to which they are harvested could have major impacts on forest ecology and wildlife habitat [11,29-31]. Residues can be used to produce some forms of bioenergy such as biofuels, but only certain types of residues such as small trees can be used to produce wood pellets. As a result, the degree of harvest of residues is a major uncertainty in any future production of forest biomass, and may range from no residue use to over 50% of residues used. Therefore, to ensure we were within the potential range of residue use, our two biomass scenarios, respectively, assumed that either 20% or 40% of forest residues would be removed.

In our two biomass scenarios, production varied by region of the state (Figure 1). Because of the existing forestry infrastructure in the Coastal Plain, we assumed that region is likely to have the greatest capacity to increase production, and we assumed that 1.36 million green tonnes (1.5 million green short tons) would be produced annually in each of the two Coastal Plain regions. The Piedmont region has some infrastructure, and thus has the capacity to increase production a moderate amount, and we assumed 0.9 green tonnes (1.0 million green short tons) of annual biomass production there. We expect no increase in forest biomass production in the Mountains because little forest harvest is done in that region.

2.3. State-and-transition simulation model and pathways (ST-Sim)

To develop state-and-transition pathways and run the STSM, we used ST-Sim software, version 2.3.0 [32]. ST-Sim is similar to, and builds upon, other STSMs, including VDDT and Path [15]. In ST-Sim, a pathway is developed for each vegetation or land-use type (see Figure 2 for an example), which defines discrete "state classes" and transitions among state classes. State classes are defined for each vegetation or land-use type based on characteristics such as successional stage and canopy cover. Transitions include ecological succession; disturbances such as wildfire; forest management, including thinning and harvest; and conversion from one land-use or vegetation type to another. Initial conditions define the starting point for the model, and the simulation proceeds from the initial landscape on an annual time step. At each time step, transitions among state classes occur in a semi-Markov framework, based on rules and probabilities set in the state-and-transition models. Limits to the total amount of forest management and land conversion in any given time step are set by time series called "transition targets" that come from SRTS outputs. Therefore, there are three main types of inputs to ST-Sim: (1) state-and-transition model pathways for each of the vegetation or land-use types in NC, (2) a table describing initial landscape conditions, including the areas of NC in each state class for all vegetation and land-use types, and (3) time series of annual areas to be thinned, harvested, or converted, from SRTS model output. We discuss (1) below, and discuss (2) and (3) in subsequent sections. ST-Sim can be run spatially or aspatially. Here we are running the model aspatially because we are focused on linking the SRTS and ST-Sim models, and examining differences among scenarios in terms of changes in areal extents of vegetation and land use types.

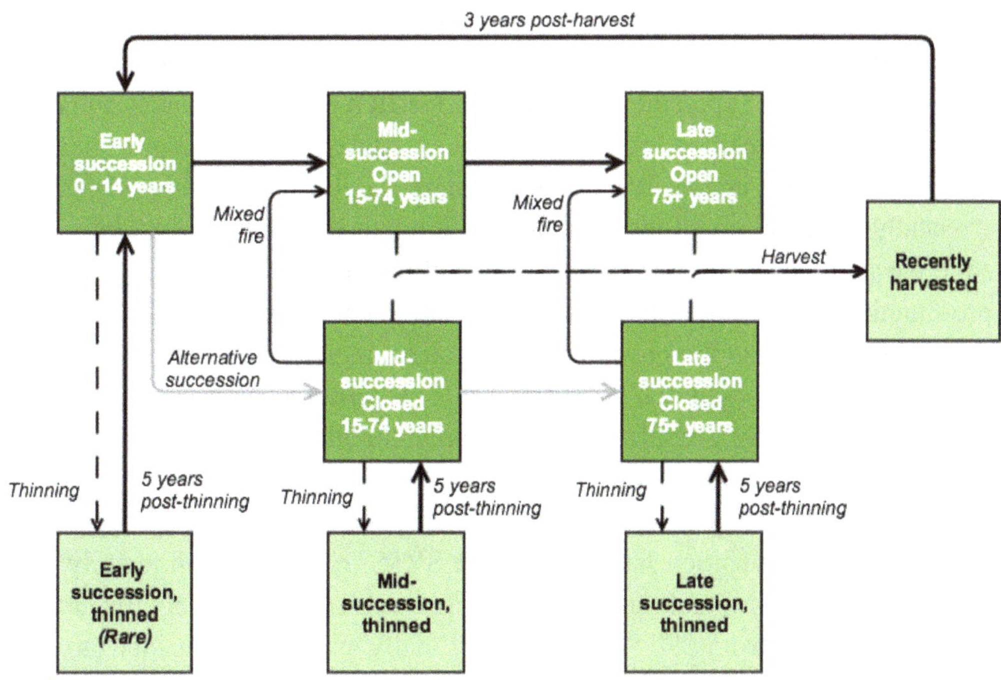

Figure 2. Example state-and-transition model pathway for Atlantic Coastal Plain Upland Longleaf Pine Woodland. Dark boxes indicate state classes from the original LANDFIRE model, and light boxes are those added for the current study. In addition to the transitions illustrated, all state classes can be converted to urban land use. All state classes except recently-harvested and recently-thinned can also experience a surface fire, which keeps the state class unchanged, and a replacement fire, which transitions to early succession. See supplementary material for transition probabilities.

We developed state-and-transition model pathways for 49 vegetation and land-use types in NC (Table 1). For most vegetation types, the pathways we used were modified from those developed for the LANDFIRE project (available at http://www.landfire.gov) [33]. In those pathways, states are defined for each vegetation type in terms of their successional stage (early, mid- or late succession), and their canopy cover (open or closed canopy). Transitions among state classes include ecological succession, which is deterministic after a pre-defined number of time steps, and disturbance events, including wildfires, which are probabilistic. In most cases, one LANDFIRE model pathway represented a single ecological system, as defined by NatureServe [34], but in some cases for wetland vegetation, a LANDFIRE model pathway represented a group of ecological systems. The LANDFIRE models reflect successional and disturbance dynamics in those systems that were likely present prior to European settlement. Each pathway is accompanied by written documentation describing the structural characteristics of each state class, including the canopy cover for each state class. A few other ecological systems occurred in NC but were not included here. These were systems that occurred in small patches, or were naturally barren or sparsely vegetated and were therefore not expected to contribute to landscape dynamics.

We modified the LANDFIRE model pathways in several important ways to reflect current conditions and forest biomass production scenarios. First, we reduced wildfire probabilities for every vegetation type by 93% to reflect contemporary fire suppression in the Southeast following our other recent work [35] by adding a multiplier of 0.07 for every wildfire transition in all pathways. For models of forested vegetation types, we added states and transitions to represent harvest and thinning (See Figure 2 for an example, and Supplementary Material for all model state classes and transitions). In those pathways, a thinning transition was added to every state class that had a closed canopy condition or represented early succession. We also added state classes that represented early, mid-, or late-succession "recently-thinned" conditions. Stands undergoing thinning in a given successional stage transitioned to the analogous "recently thinned" state class, in which canopy cover was reduced by 50% from pre-thinning conditions. Stands stayed in this state class for five years following thinning before transitioning back to the pre-thinned state class. In addition, a harvest transition was added to every state class, which could convert a stand to a "recently harvested" state class. Within two years of harvest, a stand was eligible to be converted to planted pine forest or row crop; otherwise, after three years, the stand transitioned back to the early succession state class of the same forest type it previously occupied. For every forest type, annual areas undergoing harvest and thinning were constrained with transition targets from the SRTS model, and while early succession state classes were eligible for harvest and thinning, these transitions rarely occurred during simulation due to age constraints from SRTS (see below). Finally, we added transitions to every state class in every modified LANDFIRE model pathway to represent conversion to urbanization (see below for probabilities).

In addition to modifying LANDFIRE model pathways, we created model pathways to reflect anthropogenic land uses that were not modeled by LANDFIRE, including planted pine, urban land, row crop, and pasture (see Supplementary Material for all model state classes and transitions). All of the new pathways for anthropogenic land uses consisted of a single state class, except planted pine. The planted pine model pathway consisted of two portions, a "managed" portion and a "non-stocked" portion. The planted portion represented stands that had been planted and actively managed for forest products (either conventional uses such as timber, or biomass). The non-stocked portion represented areas that had been planted and managed for forest products in the past, or were in agriculture, but were naturally regenerating after harvest. In addition, conversion of agricultural lands to natural vegetation transitioned to the early succession state class of the non-stocked model. As mentioned above, the recently harvested state class in every forest vegetation type was eligible to be converted to row crop.

Finally, for all state classes in every vegetation and land-use model pathway, we added a transition to urbanization, with probabilities that varied by region according to a summary of recent urbanization projections [24] (0.0023 for state classes in the Southern Coastal Plain, 0.0012 for the Northern Coastal Plain, 0.0152 for the Piedmont, and 0.0079 for the Mountains).

2.4. Initial conditions

The initial condition file required for using the state-and-transition models in NC was a table of the areas of each land cover class, along with areas and ages (in years) by vegetation type and state class. To develop this table, we used recent land cover and other environmental spatial data. The Southeast GAP 2001 land cover map [26] was used to distinguish the 49 land cover and vegetation types. The vegetation types for which we did not develop state-and-transition model pathways were not included. The extent of urban lands was updated to reflect 2010 conditions based on recent data [24]. Within each vegetation type, we further distinguished state classes based on canopy cover and successional stage. We used LANDFIRE 2008 succession class data (s-class) [36] to assign a successional stage and canopy cover (early mid, or late succession; open or closed canopy) for most vegetation types. However, inspection of the s-class data for the Southern Piedmont Dry Oak-Pine Forest vegetation type, which made up 20.6% of the initial mapped area of NC (Table 1), indicated that the open canopy condition was overrepresented in the data for the mid-succession stage of that vegetation type. The majority of the mid-succession stage was classified as open canopy in the s-class data, while in reality most oak-pine forests in the Piedmont have a relatively closed canopy. Therefore, we used canopy cover data from the National Land Cover Database [37] to distinguish open and closed canopy states for that vegetation type, according to the LANDFIRE model description [38]. For all vegetation types, we assigned ages to match the distribution of ages in forest types according to data from the USDA Forest Service Forest Inventory and Analysis (FIA) database [39]. Here, we used data from the FIA survey for forest plots in NC that was closest to our initial year (2010).

To compile our initial conditions input into ST-Sim, for each of the four modeled regions of NC, we summarized the areas of vegetation types by successional stage, canopy cover, and age. We also summarized areas of each urban and agriculture land-use type. We assumed lands under protection for conservation would not be subject to management for biomass, so we used the U.S. Protected Areas Database [23] to exclude those lands. See Supplementary Material for ST-Sim initial conditions. For SRTS modeling, we further summarized the land uses, vegetation types and ages using a crosswalk with five general forest types, agriculture, and urban land (see text below, and Table 1). Non-forest vegetation such as shrubland, wetland, and grassland was not included in SRTS.

2.5. Timber supply modeling (SRTS)

In order to simulate forest management and land-use change under alternative scenarios of forest biomass production in ST-Sim, we required time series of areal amounts of forest types undergoing management (harvest or thinning) and land-use conversion. Determining the effects of biomass demand on forests requires demand assumptions for the full range of forest products, including traditional small (pulpwood) and large (sawlogs) roundwood. Except for logging residues, these products will compete with bioenergy. The resulting price effects can lead to “displacement” of existing uses or “leakage” of demand to other regions. Therefore, the effects of biomass production on land use and forest management (thinning and harvesting) will not be additive. To incorporate that non-additivity, or “displacement” of other forest products, as well as baseline production, we used

SRTS, which has been used in assessments of forest resources in the South and beyond [21,22]

The SRTS model operates on an annual time step and combines economic resource allocation with biological growth to link timber markets (price and harvest for a range of timber products) with forest dynamics. At each time step, the model simulates the impact of exogenous demand scenarios on the areas of general forest types in 5-year age classes that are harvested or thinned, along with the total areas of those forest types, and total areas of agricultural and urban land. For each region (group of counties) modeled, the forest inventory is adjusted annually from the initial conditions by adding net growth and subtracting the estimated harvest. The interactions between forest and agricultural land is driven by land rents, which include net income from forest harvest [14,40]. In SRTS, annual areas of each forest and land-use type are determined based on an index of timber prices from the endogenous market model combined with a response surface of agricultural and forest land prices from a recent land-use model [41] updated with the most recent land use data [42]. In this study we used recent urbanization projections [24] to define transitions from rural to urban land. Rural land is allocated between agricultural and forest land based on relative economic returns each year.

For SRTS modeling, we used the same initial conditions file described above, but aggregated vegetation types into general forest types, including one planted pine type and four natural forest types, and 5-year age classes. Because SRTS models forest change only, some vegetation types (such as marshes and beach vegetation) were excluded from SRTS modeling. In addition, some forest vegetation types were also excluded from SRTS because they are unlikely to undergo intensive management for biomass. Such is the case for maritime forest types, which exist in small patches on sandy soils, and have low productivity. This meant that harvest, thinning, and agricultural and urban land-use conversion for these vegetation types were not included in SRTS or ST-Sim modeling. We believe this is reasonable because these vegetation types are not likely to be used for biomass. In addition, many of these types are protected from land conversion (e.g. beach dunes) or are wetland vegetation, which has a low probability of conversion in the urbanization model [24].

When running SRTS models, we assumed in all scenarios that empirical demand for all forest products except pine sawtimber would increase 0.5% per year over recent FIA plot data throughout the simulation period. Because the recent recession greatly affected housing starts, and pine sawtimber is used for house construction, we adjusted pine sawtimber during the initial years of the simulation to reflect recession recovery. Pine sawtimber demand was increased by 3% per year from 2011–2014, by 1% per year from 2015–2019, and by 0.5% per year thereafter. All else was held constant in the baseline scenario. For the two scenarios that included forest biomass, we wanted to simulate the effects of an increase in biomass in the relatively near future, but realized that demand would take some time to reach target levels. Therefore, we increased demand gradually in these scenarios starting in 2015 until it reached the target demand in 2019 (Figure 1). Demand was then held constant until the end of the projection. In these scenarios for a given region, we assumed the forest types used for biomass and forest logging residues would be the same as the mix of forest trees harvested. Depending on the scenario, up to 20% or 40% of available logging residues were used for both hardwood and pine forest types. The result of SRTS for each region and modeled scenario included two types of annual time series projections: (1) the areas thinned and harvested for the five general forest types and by age class, and (2) the total areas in natural forest, planted pine forest, agriculture, and urban land areas over time. These were used as input to ST-Sim.

2.6. Linking SRTS output with ST-Sim and running state-and-transition simulations

Before input into ST-Sim, the two SRTS outputs for each scenario and region modeled were manipulated to produce annual time series transition targets for land-use conversion and forest management. For the first output, annual areas of thinning and harvest by forest type and age class, the time series were smoothed by averaging over a five-year moving window because SRTS output values showed a high level of interannual variation, which is likely unrealistic.

We converted the second SRTS output, total areas of forest and land-use type by year, into annual transition targets for conversion among land-use types, with the exception of conversion of all types to urbanization, which we had data for from a previous study [24]. To do so, for a given land-use or forest type, we calculated the difference between the total area in a given year and the area for the previous year. We then removed any conversion to urban in the SRTS output by assuming that loss of all land-use types to urban from one year to the next was proportional to the relative areas of those land-use types. Thus, for example, if twice as much forest existed as agriculture, forest loss contributed twice as much area to urbanization as agriculture contributed. We then used a similar approach to compare the remaining (non-urbanization) changes in forest types and agriculture. We thus calculated a set of time series of areas converted among agriculture, planted pine forest, and natural forest types. We also averaged these time series over a five-year moving window. The result was transition targets for conversion among natural forest, and agricultural land uses. Note that because SRTS output provides annual areas that remain in each land-use type each year, the resulting transition targets capture only the net changes among land-use types for a given region. In reality, the annual areas in each land-use type are the result of conversion to and from each type. For example, a net increase in planted pine forest may be the result of conversion from agriculture, but in reality some conversion of planted pine to agriculture may have also occurred. We do not capture these gross changes.

The transition targets were applied to transitions in ST-Sim based on the crosswalk between the ST-Sim and SRTS land-use and vegetation types (Table 1), and were constrained to state classes with the appropriate ages. For example, a transition target area for thinning in mixed pine hardwood forest, ages 16–20, in the Piedmont was applied to all ST-Sim vegetation types in that region that corresponded to mixed pine hardwood forest. Further, within those ST-Sim vegetation types, the targets were applied to state classes that correspond to ages 16-20. If multiple state classes overlapped the age range (i.e., if in this case, early succession lasted until age 17, and mid-succession began at age 18), the transition target was applied to both state classes. This meant that a single SRTS transition target was often applied to multiple ST-Sim vegetation types or state classes. The ST-Sim software accommodates this by distributing transition targets among systems based on the proportion of each vegetation type and age class on the landscape in a given time step. For example, if a target is to be split between two ST-Sim state classes, and one has twice as much area as the other, the total area thinned is divided in a ratio of 2:1 to meet the transition target for the given year.

2.7. Simulating landscape dynamics

We simulated changes within and among vegetation and land-use types over time in all four regions of NC using ST-Sim under the three future scenarios from 2011–2050. For simulation, ST-Sim requires a total simulation area and number of simulation cells. We set the total simulation area to equal the total area in the initial conditions in each region, and set the number of simulation cells to be such that the cell size was approximately 10 ha. In other words, the number of cells was set to total landscape size (in ha)/10 ha. For each scenario and region, we ran 50 Monte Carlo iterations of the ST-Sim model. We chose 50 iterations to ensure outputs reflected the range of possible outcomes due to the stochastic aspects of ST-Sim, while maintaining reasonable model run times.

To address our first objective regarding linking ST-Sim and SRTS, for land-use transitions, harvest, and thinning, we examined the correspondence between annual SRTS targets matched annual transition areas in ST-Sim output. To address our second objective, we compared dynamics of major land uses and general vegetation types between the two bioenergy scenarios, and compared each to the baseline scenario. We illustrate differences in vegetation dynamics among scenarios using two example vegetation types: the Atlantic Coastal Plain Upland Longleaf Pine Woodland and planted pine forest.

3. Results

We modeled landscape dynamics across 9,992,735 hectares (999,274 km^2), or 79% of the land area in NC. Lands that were barren or sparsely vegetated, as well as public lands, were not included in our simulations. The comparison of transition targets derived from SRTS output for land-use change, harvest, and thinning to ST-Sim modeled transitions indicated relatively close agreement for most transition types and scenarios (Table 2). The annual area undergoing each land-use change transition generally differed by 5% or less from transition targets on average. In many cases, the difference was less than 2%. Some harvest and thinning transitions showed slightly higher average differences, with three transitions in planted pine forest types averaging more than 8% difference from transition targets. However, most harvest and thinning transitions also differed by less than 2% from annual targets.

Landscape dynamics, including areas of land-use and vegetation types on the landscape, as well as transitions due to thinning, harvest, and land-use change, showed some variation among the scenarios we simulated. In each case, dynamics were similar among scenarios up to the year 2015, when we added biomass production for bioenergy. In all scenarios over the simulation period across NC, the total areas of forest and agriculture decreased, while urban land increased (Figure 3). This result is not surprising, given the urbanization projections that were incorporated into these scenarios. However, there were important differences in these trends among scenarios. By the end of the simulation, the baseline scenario had the smallest area of forest and the largest area of agriculture remaining. Compared to the other two scenarios, the biomass scenario that incorporated 20% forest residue removal resulted in more forest and less agriculture on the landscape by the end of the period. The biomass scenario that incorporated 40% removal of forest residues resulted in an intermediate amount of agriculture and forest on the landscape compared with the other two scenarios, and the results were qualitatively similar to the baseline scenario. Because we simulated dynamics over time for over 200 state classes in all vegetation and land-use types, we are unable to show detailed results by state class over time for each, but we present the dynamics of change for two vegetation types below.

Overall, across the simulation period, the areas of forest thinned and harvested differed among scenarios (Figure 4). After the first few years of the simulation in which biomass production was gradually increased, there was a smaller area of forest harvested and thinned annually under the baseline scenario than under the two biomass scenarios. The biomass scenario that assumed only 20%

of forest residues were removed resulted in the highest annual areas of thinning and harvest for most time steps. The scenario that assumed 40% residue removal resulted in areas of thinning and harvest that were intermediate between the other two scenarios for most time steps. There were some time steps in which either harvest or thinning amounts were highest for the scenario that incorporated 40% of residues, but there were no time steps in which both harvest and thinning were greater for the 40% residue scenario.

Trends in land-use change transitions across NC over time were similar for the baseline and 40% residue scenario, but differed slightly for the scenario that assumed 20% of residues (Figure 5). In particular, for most time steps in the simulation, the 20% residue scenario showed a larger area converted from agriculture to planted pine and natural forest than the other two scenarios. For a portion of the simulation period, that scenario also showed a higher area of natural forest converted to agriculture.

Table 2. Average area (ha) undergoing harvest, thinning, and land-use transitions[a] in ST-Sim, and the average difference between ST-Sim output and targets developed from SRTS.

Transition	Baseline		Biomass, 20% residues		Biomass, 40% residues	
	ST-Sim avg. area	Difference (% of avg.)	ST-Sim avg. area	Difference (% of total)	ST-Sim avg. area	Difference (% of total)
Harvest, planted pine	20,534	2,744 (13.4%)	20,976	1001 (4.8%)	19,550	1,668 (8.5%)
Harvest, natural pine	16,125	129 (0.8%)	18,674	126 (0.7%)	20,126	137 (0.7%)
Harvest, mixed pine hardwood	17,131	1,126 (6.6%)	23,972	1298 (5.4%)	19,255	1,178 (6.1%)
Harvest, upland hardwood	16,933	68 (0.4%)	19,026	66 (0.3%)	17,184	72 (0.4%)
Harvest, lowland hardwood	9,354	31 (0.3%)	13,903	50 (0.4%)	12,652	43 (0.3%)
Thinning, planted pine	35,654	3,347 (9.4%)	46,630	717 (1.5%)	41,895	1,795 (4.3%)
Thinning, natural pine	15,086	66 (0.4%)	14,332	67 (0.5%)	15,250	58 (0.4%)
Thinning, mixed pine hardwood	31,051	1,750 (5.6%)	30,136	1762 (5.8%)	29,675	1,741 (5.9%)
Thinning, upland hardwood	10,140	37 (0.4%)	9,504	35 (0.4%)	11,096	37 (0.3%)
Thinning, lowland hardwood	6,638	33 (0.5%)	8,140	37 (0.5%)	7,505	33 (0.4%)
Ag to natural forest	7,984	29 (0.4%)	9,232	31 (0.3%)	8,241	25 (0.3%)
Ag to planted pine	2,521	14 (0.6%)	5,568	25 (0.4%)	2,818	17 (0.6%)
Natural forest to ag	1,984	100 (5.0%)	2,550	108 (4.2%)	1,967	101 (5.1%)
Planted pine to ag	2,143	27 (1.3%)	1,980	23 (1.2%)	2,157	25 (1.1%)
Planted pine to natural forest	398	7 (1.8%)	219	7 (3.1%)	345	5 (1.4%)

[a] Although natural forest to planted pine transitions were possible in SRTS and ST-Sim, they did not occur.

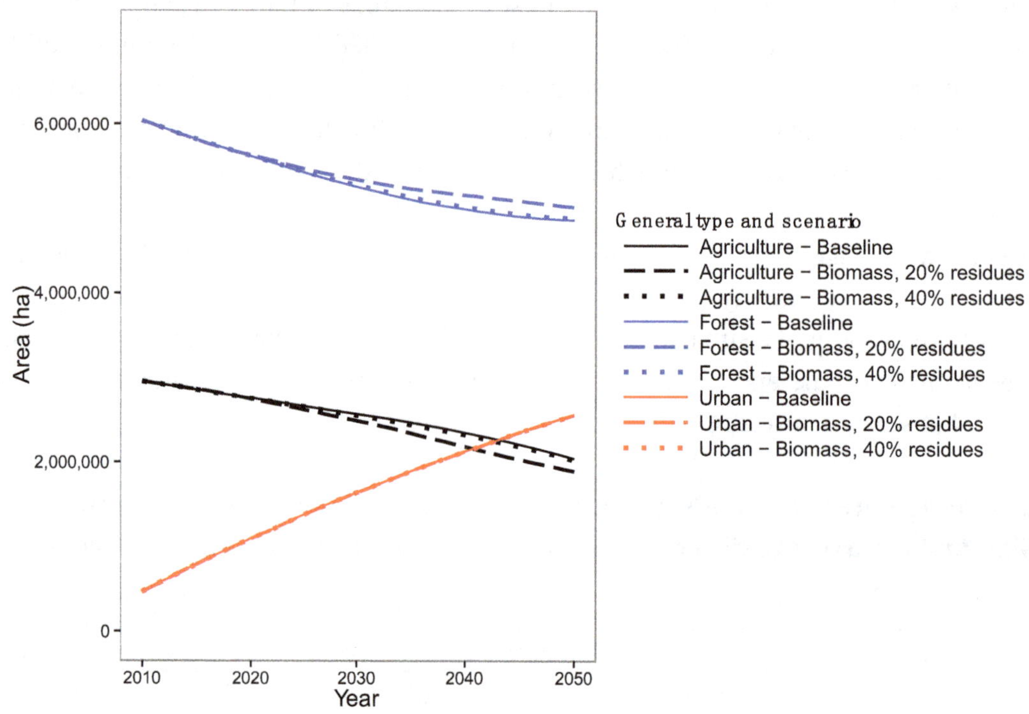

Figure 3. Total areas of three general land-use types over time: forest, agriculture, and urban. Lines represent means of 50 Monte Carlo simulations. The range of results across the 50 simulations is too narrow to show up on this graph. Forest vegetation and agriculture types were aggregated according to Table 1. Wetland and grass/shrub vegetation were also modeled but make up only a small portion of NC (Table 1) and showed little change in area over time.

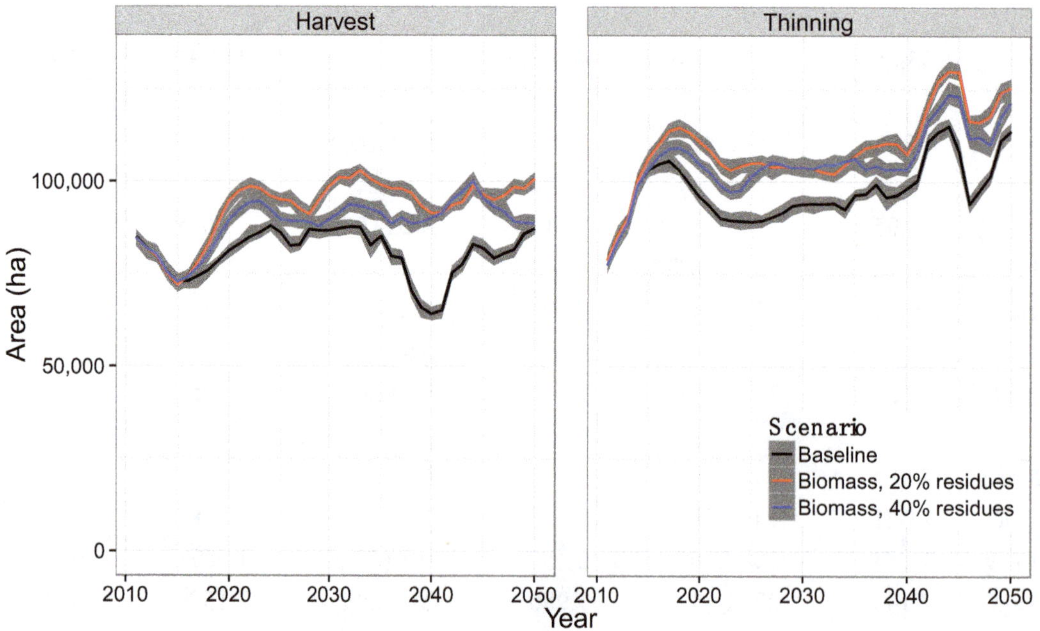

Figure 4. Annual areas of forest harvest and thinning in ST-Sim output across NC for the three modeled scenarios. Gray shading indicates the range of 50 Monte Carlo simulations.

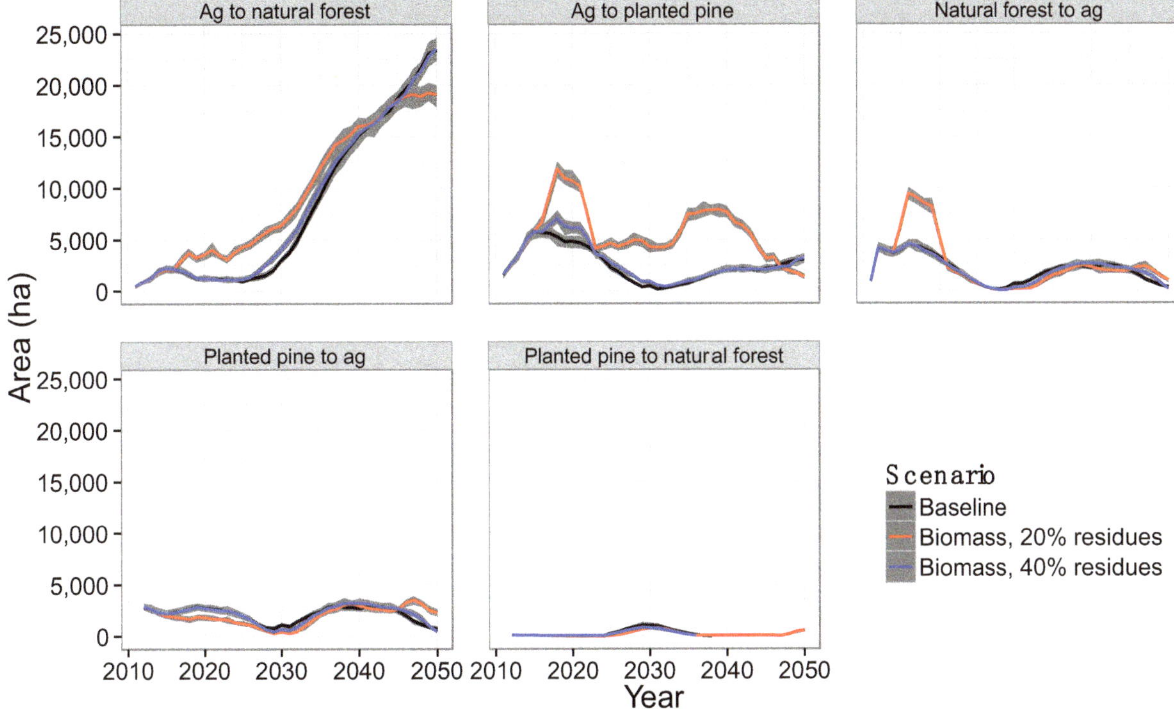

Figure 5. Annual areas of land-use change transitions in ST-Sim output across NC for the three modeled scenarios. Gray shading indicates the range of 50 Monte Carlo simulations. One other transition, natural forest to planted pine, was possible but had a value of 0 in transition targets and ST-Sim output. Note that for any single region in a given year, a transition in only one direction is possible (i.e., from agriculture to natural forest, but not the reverse), but this shows aggregate output for four regions.

Landscape dynamics results for the Atlantic Coastal Plain Upland Longleaf Pine Woodland were relatively similar among all scenarios (Figure 6). Areas in the early succession and mid-succession open canopy state classes decreased over time, while the late-succession closed canopy state class increased in area. Areas in the late-succession open and thinned state classes remained even over time, while the mid-succession thinned state class increased in area for the first five years, but then decreased. The only substantial differences among scenarios were in the recently harvested and mid-succession closed canopy states. The baseline scenario showed more area in mid-succession closed canopy and less in the recently harvested state compared with the other two scenarios in some years, due to the lower harvest rates for natural pine forest in that scenario (Table 2).

The dynamics of the state classes within the planted pine forest vegetation type over time show differences among scenarios (Figure 7). Compared to the baseline scenario, planted pine forests under the two biomass scenarios have less area in the two mid-succession closed canopy state classes, and a greater area in both late succession state classes. In addition, compared with the baseline scenario for most time steps in the simulation, the two biomass scenarios show a greater area of planted pine forest the mid- and late-succession recently thinned classes. The three scenarios show approximately the same areas in early succession state classes over time, but the biomass scenario that incorporates 20% residue removal has a slightly greater area than the other two scenarios.

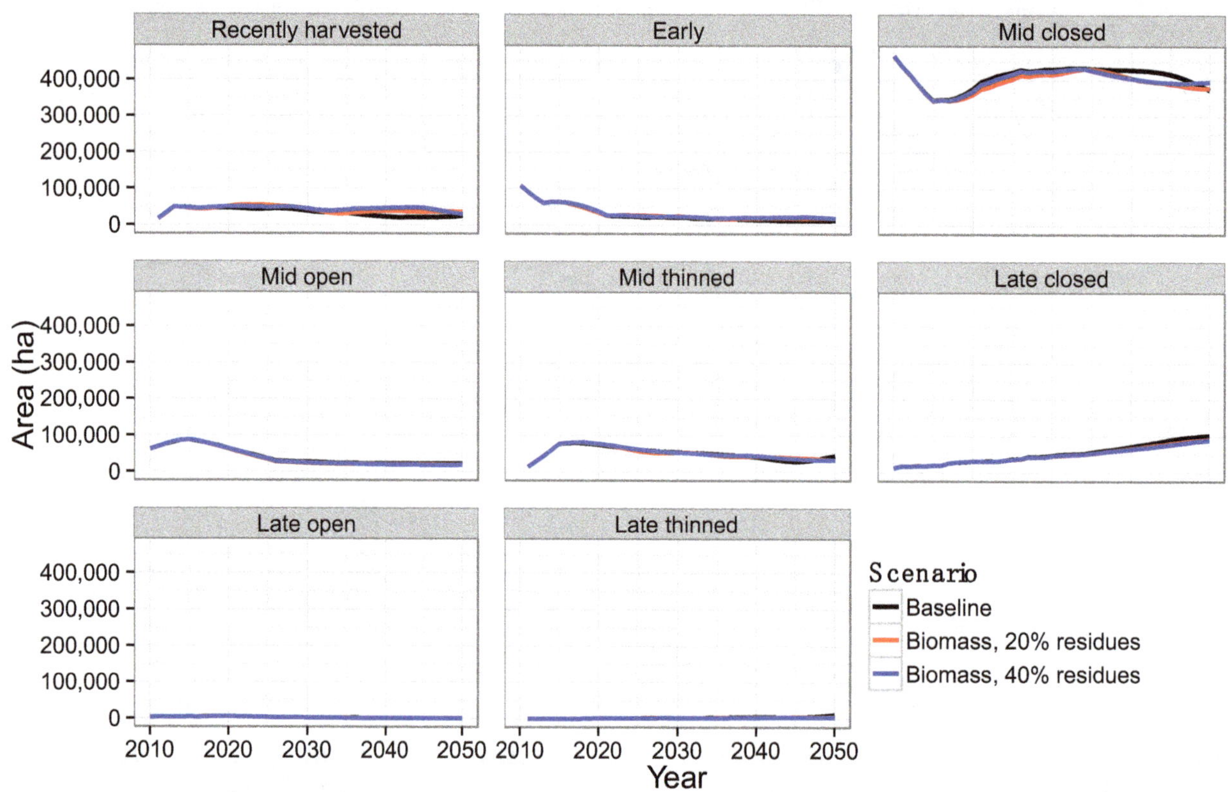

Figure 6. Areas of each state class over time for Atlantic Coastal Plain Upland Longleaf Pine Woodland across its range in NC.

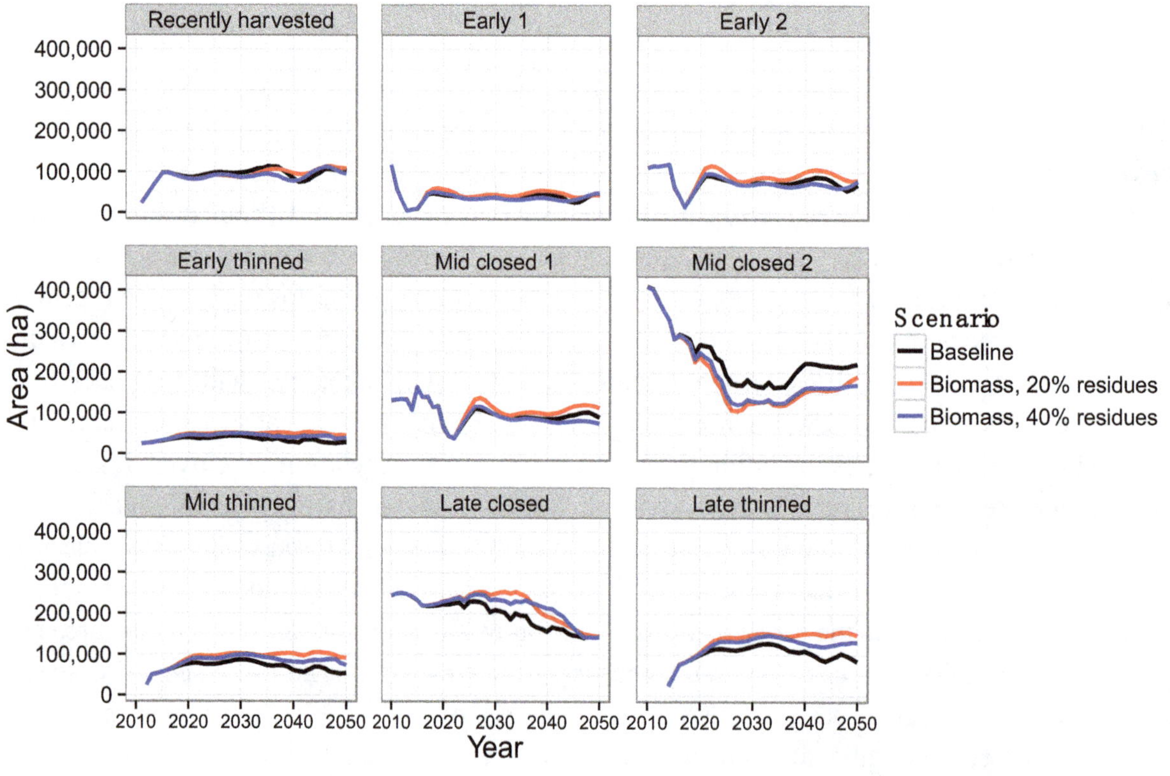

Figure 7. Areas of each state class over time for planted pine forest across its range in NC.

4. Discussion

The future impacts and sustainability of bioenergy production on landscapes are uncertain, but depend on landscape dynamics [43]. To investigate the potential landscape impacts of bioenergy production, we simulated landscape dynamics under alternative forest biomass production scenarios in North Carolina. Results show that targets for land-use conversions as well as forest harvest and thinning under alternative scenarios produced by the economics based timber supply model SRTS can be used to inform landscape dynamics simulation in the state-and-transition simulation model (STSM) ST-Sim. The scenarios we modeled resulted in key differences in vegetation and land-use changes. While biomass scenarios led to more forest on the landscape, the scenarios also led to more recently harvested and thinned forest, and less closed canopy forest.

In most cases, areas undergoing forest and land-use transitions in the ST-Sim model output corresponded fairly well with transition targets set by SRTS. For all land-use change conversions, the ST-Sim model produced transition areas close to the SRTS target. In a few cases, for harvest and thinning, especially for planted pine forest, the ST-Sim transitions were different from the SRTS targets. Where these differences occurred, we believe they were likely due to differences between the two models in how they represent forest dynamics. For example, in SRTS, all land in each forest type is available to be thinned at any point in time, but in ST-Sim state-and-transition models, only the portion of forest in the closed canopy or early succession state classes was able to be thinned. Thus, compared with SRTS, the available forest area for thinning in any given year in ST-Sim could be lower, and target areas for thinning may not always be achievable. Because these differences were relatively minor in any given year, they would likely make little difference to overall forest dynamics and production of forest products and biomass. But, more tightly linking SRTS and ST-Sim could help ensure better agreement between the two models and facilitate ease of use. For example, dynamic coupling of the models so that they run simultaneously, so that for each time step, SRTS output serves as transition targets in ST-Sim, and areas in each forest vegetation type, state class, and land-use class from ST-Sim are aggregated and input back into SRTS for the following time step.

A comparison of general trends in land use among the three biomass scenarios show that while urban land area increased at the expense of forest and agricultural land in all scenarios, the amounts of agricultural and forest land remaining by the end of the simulation period varied by scenario: the two biomass scenarios resulted in more forest and less agricultural land than the baseline scenario. The increased demand for forest products under the two biomass scenarios resulted in more conversion of forest to agriculture, and less conversion of agriculture to forest. This result is consistent with other studies that stress the importance of strong forest product markets for retaining forests on the landscape in the Southeast [13].

However, by enabling an examination of the trends within individual forest vegetation types, this study extends previous studies. Results indicate that increased harvest and thinning under the two biomass scenarios resulted in more intensively-managed forest on the landscape. Planted pine forest dynamics show more recently harvested and thinned forest, and less closed canopy forest on the landscape by the end of the simulation period under the biomass scenarios than under the baseline scenario. While differences were smaller for the Longleaf Pine Woodland, dynamics for that vegetation type also showed a slight increase in recently harvested forest at the expense of closed canopy forest. Therefore, while biomass scenarios may keep more forest on the landscape than a business-as-usual future, that forest may be more intensively managed.

This study also extends previous studies by examining landscape impacts of alternative residue removal scenarios. Differences between the two biomass scenarios show that assuming a higher rate of forest residue removal during harvest resulted in smaller areas of thinning, harvest, and some land-use changes than a lower residue removal rate. Therefore, it is clear that forest residues were able to meet some of the demand for biomass. In some cases, for some types of transitions such as conversion of agriculture to planted pine, the scenario with lower residue removal resulted in dynamics that were almost identical to the baseline scenario. This result suggests that an even higher rate of residue removal, such as 70–80%, as recommended by some guidelines for liquid biofuels [29,30], may not cause much, if any, difference in landscape dynamics compared with a baseline, non-bioenergy scenario.

A strength of this modeling effort linking SRTS and ST-Sim is the ability to incorporate temporal variation in annual areas of harvest and thinning. For example, toward the end of the simulation period, the area thinned under all scenarios decreased and then increased. Area harvested followed a similar, but opposite, trend for the same time period. These fluctuations reflect two aspects of the models. First, the SRTS and ST-Sim models both incorporate varying eligibility for harvest and thinning due to forest age. If a large area of forests is of a similar age and therefore becomes eligible for thinning or harvest all at once, fluctuations in harvest or thinning could be pronounced. In addition, harvest and thinning combine to meet the biomass targets in SRTS. So, when one of these management transitions decreases, an increase in the other is likely. By incorporating these temporal variations, we are able to go beyond recent modeling efforts that have assessed potential bioenergy impacts for the current landscape only [44].

While our scenarios provide insight into landscape-level changes due to forest biomass production, the local ecological effects of increased management intensity, including forest residue removal, under biomass scenarios may be positive or negative, depending on the ecological target, and should be explored further. For example, while removing forest residues lessens landscape effects, the practice may have negative local effects on amphibian species that nest in snags or down logs in some vegetation types [45], though effects may vary by species and vegetation type [46,47]. In addition, forest harvest and thinning may have positive effects on some bird species that use early successional habitats or forests with reduced tree canopies but not on other species [47,48,49]. Furthermore, North Carolina contains several vegetation types with high diversity. Woodlands that make up the longleaf pine ecosystem have among the highest levels of plant richness per square meter of any ecosystem in the world [51] and the North American Coastal Plain, of which North Carolina's Coastal Plain is a part, is recognized as a global biodiversity hotspot [25]. Therefore, any changes due to bioenergy production that result in forest conversion or more intense management have the potential to impact this biodiversity.

One caveat to our simulation results is that our models have not been fully validated by conducting historical simulations and testing outputs against observations. Full validation and model testing should be an important next step in the development of our modeling framework. However, our objective in this paper is to report on differences among biomass scenarios, which depend on scenarios of land conversion and management from SRTS. Importantly, SRTS is based on empirical observations from USDA Forest Service Forest Inventory and Analysis data. Therefore, we are confident in our conclusions about differences among the scenarios we simulated.

In addition to the further research already discussed, the STSMs presented here could be expanded to incorporate additional scenarios and realism. For example, the proximity of sites to an existing or planned facility that processes biomass into bioenergy may affect the likelihood of biomass harvest. Spatially explicit STSMs would allow the incorporation of mapped suitability surfaces for changes in production depending on geography or site characteristics. Furthermore, incorporating additional feedstocks for bioenergy, including so-called "purpose-grown crops" for biofuels, such as switchgrass and sweet sorghum, would allow investigation of the effects of a wider range of production scenarios.

5. Conclusion

Taken together, our results show the importance and utility of linking a forest economics model with an STSM to investigate landscape dynamics in response to alternative biomass production scenarios. The non-linear and variable trends of forest management and land-use conversions through time due to the interactions between successional dynamics, forest product markets, and demand for forest land would be difficult to determine without SRTS output. And the use of STSMs allowed us to investigate the dynamics of specific vegetation types and state classes, which other studies based on field data or static landscapes have not addressed.

Overall, landscape dynamics results show that compared with the baseline scenario, forest biomass production scenarios lead to more forest and, specifically, more intensively managed forest on the landscape by the end of the simulation. However, even a moderate harvest rate for forest residues dampens the effects of biomass production on landscape dynamics. These results suggest that on one hand, biomass production for bioenergy may have some landscape effects that could negatively impact species and ecosystems in the state. On the other hand, those impacts could be minimized by incorporating certain harvesting practices. Therefore, this research points to a future in which biomass has the potential to be produced sustainably, with minimal impacts to landscapes.

Acknowledgments

We thank L. Frid and C. Daniel of ApexRMS for assistance with development of state-and-transition models and simulation in ST-Sim. T. Earnhardt helped develop initial conditions, and R. Gonzalez assisted with biomass to biofuel conversion. We also thank A. Costanza for help with model processing. Funding for this work was provided by the USGS Gap Analysis Program and the Biofuels Center of North Carolina. Any use of trade, product, or firms names is for descriptive purposes only and does not imply endorsement by the U.S. Government.

Conflict of Interest

All authors declare no conflicts of interest in this paper.

Supplementary

1. Biomass to ethanol conversion steps.
2. Full list of deterministic transitions for all state-and-transition models.
3. Full list of probabilistic transitions for all state-and-transition models.
4. List of all initial conditions used as one input into ST-Sim.

References

1. Sedjo RA, Sohngen B (2013) Wood as a Major Feedstock for Biofuel Production in the United States: Impacts on Forests and International Trade. *J Sustain For* 32: 195-211.
2. Goh CS, Junginger M, Cocchi M, et al. (2013) Wood pellet market and trade: a global perspective. *Biofuels Bioprod Bioref* 7: 24-42.
3. Immerzeel DJ, Verweij PA, van der Hilst F, et al. (2014) Biodiversity impacts of bioenergy crop production: A state-of-the-art review. *GCB Bioenergy* 6: 183-209.
4. McDonald RI, Fargione J, Kiesecker J, et al. (2009) Energy sprawl or energy efficiency: Climate policy impacts on natural habitat for the United States of America. *PLoS One* 4: e6802.
5. Stoms DM, Davis FW, Jenner MW, et al. (2012) Modeling wildlife and other trade-offs with biofuel crop production. *GCB Bioenergy* 4: 330-341.
6. Dale VH, Kline KL, Wiens J, et al. (2010) Biofuels : Implications for Land Use and Biodiversity. The Ecological Society of America Biofuels and Sustainability Reports. Available from: http://www.esa.org/biofuelsreports/files/ESA%20Biofuels%20Report_VH%20Dale%20et%20al.pdf.
7. Wiens J, Fargione J, Hill J (2011) Biofuels and biodiversity. *Ecol Appl* 21: 1085-1095.
8. Daystar J (2014) Environmental Impacts of Cellulosic Biofuels Made in the South East: Implications of Impact Assessment Methods and Study Assumptions. North Carolina State University: 264 pages.
9. Wear D, Abt R, Alavalapati J, et al. (2010) The South's Outlook for Sustainable Forest Bioenergy and Biofuels Production. The Pinchot Institute Report. Available from: http://www.pinchot.org/uploads/download?fileId=512.
10. Fletcher RJ, Robertson BA, Evans J, et al. (2011) Biodiversity conservation in the era of biofuels: risks and opportunities. *Front Ecol Environ* 9: 161-168.
11. Riffell S, Verschuyl J, Miller D, et al. (2011) Biofuel harvests, coarse woody debris, and biodiversity – A meta-analysis. *For Ecol Manage* 261: 878-887.
12. Dale VH, Lowrance R, Mulholland P, et al. (2010) Bioenergy Sustainability at the Regional Scale. *Ecol Soc* 15: 23.
13. Wear DN, Huggett R, Li R, et al. (2013) Forecasts of Forest Conditions in U.S. Regions under Future Scenarios: A Technical Document Supporting the Forest Service 2010 RPA Assessment. *Gen Tech Rep* SRS-170.
14. Lubowski RN, Plantinga AJ, Stavins RN (2008) What Drives Land-Use Change in the United States? A National Analysis of Landowner Decisions. *Land Econ* 84: 529-550.
15. Daniel CJ, Frid L (2012) Predicting Landscape Vegetation Dynamics Using State-and-Transition Simulation Models. *Proc First Landsc State-and-Transition Simul Model Conf June 14-16* 2011: 5-22.
16. Bestelmeyer BT, Herrick JE, Brown JR, et al. (2004) Land management in the American southwest: a state-and-transition approach to ecosystem complexity. *Environ Manage* 34: 38-51.
17. Costanza JK, Hulcr J, Koch FH, et al. (2012) Simulating the effects of the southern pine beetle on regional dynamics 60 years into the future. *Ecol Modell* 244: 93-103.
18. Wilson T, Costanza J, Smith J, et al. (2014) Second State-and-Transition Simulation Modeling Conference. *Bull Ecol Soc Am* 96: 174-175.

19. Halofsky J, Halofsky J, Burscu T, et al. (2014) Dry forest resilience varies under simulated climate-management scenarios in a central Oregon, USA landscape. *Ecol Appl* 24: 1908-1925.
20. Provencher L, Forbis TA, Frid L, et al. (2007) Comparing alternative management strategies of fire, grazing, and weed control using spatial modeling. *Ecol Modell* 209: 249-263.
21. Abt R, Cubbage F, Abt K (2009) Projecting southern timber supply for multiple products by subregion. *For Prod J* 59: 7-16.
22. Abt KL, Abt RC, Galik CS, et al. (2014) Effect of Policies on Pellet Production and Forests in the U.S. South: A Technical Document Supporting the Forest Service Update of the 2010 RPA Assessment. *Gen Tech Rep* GTR-SRS-202.
23. U.S. Geological Survey National Gap Analysis Program (2013) Protected Areas Database-US (PAD-US), Version 1.3. Available from: http://gapanalysis.usgs.gov/padus/.
24. Terando A, Costanza JK, Belyea C, et al. (2014) The southern megalopolis: using the past to predict the future of urban sprawl in the Southeast U.S. *PLoS One* 9: e102261.
25. Noss RF, Platt WJ, Sorrie BA, et al. (2015) How global biodiversity hotspots may go unrecognized: lessons from the North American Coastal Plain. Richardson D, ed. *Divers Distrib* 21: 236-244.
26. Southeast Gap Analysis Project (SEGAP) (2008) Southeast GAP regional land cover [digital data]. Available from: http://www.basic.ncsu.edu/segap/.
27. Burke S, Hall BR, Shahbazi G, et al. (2007) North Carolina's Strategic Plan for Biofuels Leadership. Available from: http://www.ces.ncsu.edu/fletcher/mcilab/publications/NC_Strategic_Plan_for_Biofuels_Leadership.pdf.
28. Forisk Consulting LLC (2014) Wood bioenergy US database 2013. Available by subscription.
29. Lal P, Alavalapati JRR, Marinescu M, et al. (2011) Developing Sustainability Indicators for Woody Biomass Harvesting in the United States. *J Sustain For* 30: 736-755.
30. Evans A, Perschel R, Kittler B, et al. (2010) Revised assessment of biomass harvesting and retention guidelines. *For Guild, St Fe, NM, USA*: 33.
31. Janowiak MK, Webster CR (2010) Promoting Ecological Sustainability in Woody Biomass Harvesting. *J For* 108: 16-23.
32. Apex Resource Management Solutions (2014) ST-Sim state-and-transition simulation model software. Available from: http//www.apexrms.com/stsm.
33. Rollins MG (2009) LANDFIRE: a nationally consistent vegetation, wildland fire, and fuel assessment. *Int J Wildl Fire* 18: 235-249.
34. Comer P, Faber-Langendoen D, Evans R, et al. (2003) *Ecological Systems of the United States: A Working Classification of U.S. Terrestrial Systems*. Arlington, VA, USA. NatureServe, 82 pages.
35. Costanza JK, Terando AJ, McKerrow AJ, et al. (2015) Modeling climate change, urbanization, and fire effects on Pinus palustris ecosystems of the southeastern U.S. *J Environ Manage* 151: 186-199.
36. LANDFIRE (2014) LANDFIRE 2008 (version 1.1.0) Succession Class (S-Class) Layer. U.S. Department of Interior, Geological Survey. Available from: Http://landfire.cr.usgs.gov/viewer.
37. Multi-Resolution Land Characteristics Consortium (MRLC) (2011) National Land Cover Database, USFS Tree Canopy Cartographic, 2014. Available from: http://www.mrlc.gov/nlcd11_data.php.

38. Mackie R, Mason J, Curcio G (2007) LANDFIRE biophysical setting model for Southern Piedmont Dry Oak(-Pine) Forest. Available from: http://www.landfire.gov/national_veg_models_op2.php.
39. USDA Forest Service (2012) Forest Inventory and Analysis Data. Available from: http://apps.fs.fed.us/fiadb-downloads/datamart.html.
40. Young T, Wang Y, Guess F, et al. (2015) Understanding the Characteristics of Non-industrial Private Forest Landowners Who Harvest Trees. *Small-scale For* 1-13.
41. Hardie I, Parks P, Gottleib P, et al. (2000) Responsiveness of Rural and Urban Land Uses to Land Rent Determinants in the U.S. South. *Land Econ* 76: 659.
42. USDA Natural Resources Conservation Service (2000) 1997 National Resources Inventory Data, Revised December 2000.
43. Dale VH, Kline KL, Wright LL, et al. (2011) Interactions among bioenergy feedstock choices, landscape dynamics, and land use. *Ecol Appl* 21: 1039-1054.
44. Evans JM, Fletcher RJ, Alavalapati JRR, et al. (2013) Forestry Bioenergy in the Southeast United States: Implications for Wildlife Habitat and Biodiversity. Availbale from: http://www.nwf.org/News-and-Magazines/Media-Center/Reports/Archive/2013/12-05-13-Forestry-Bioenergy-in-the-Southeast.aspx.
45. Owens AK, Moseley KR, McCay TS, et al. (2008) Amphibian and reptile community response to coarse woody debris manipulations in upland loblolly pine (Pinus taeda) forests. *For Ecol Manage* 256: 2078-2083.
46. Otto CR V, Kroll AJ, McKenny HC (2013) Amphibian response to downed wood retention in managed forests: A prospectus for future biomass harvest in North America. *For Ecol Manage* 304: 275-285.
47. Davis JC, Castleberry SB, Kilgo JC (2010) Influence of coarse woody debris on herpetofaunal communities in upland pine stands of the southeastern Coastal Plain. *For Ecol Manage* 259: 1111-1117.
48. Wood P, Sheehan J, Keyser P, et al. (2013) Cerulean Warbler: Management Guidelines for Enhancing Breeding Habitat in Appalachian Hardwood Forests. American Bird Conservancy. The Plains, VA, USA. 28 Pages.
49. Perry RW, Thill RE (2013) Long-term responses of disturbance-associated birds after different timber harvests. *For Ecol Manage* 307: 274-283.
50. Wilson MD, Watts BD (2000) Breeding bird communities in pine plantations on the coastal plain of North Carolina. *Chat* 64: 1-14.
51. Peet RK, Allard DJ (1993) Longleaf Pine Vegetation of the Southern Atlantic an Eastern Gulf Coast Regions: A Preliminary Classification. In: Hermann SM, ed. *Proceedings of the Tall Timbers Fire Ecology Conference, No. 18, The Longleaf Pine Ecosystem: Ecology, Restoration and Management*. Tallahassee, FL, USA: Tall Timbers Research Station.; 1993: 45-81.

9

Modulation of metallothionein, pi-GST and Se-GPx mRNA expression in the freshwater bivalve *Dreissena polymorpha* transplanted into polluted areas

Périne Doyen [1], Etienne Morhain [2] and François Rodius [2]

[1] Université du Littoral Côte d'Opale, Institut Charles Viollette, Equipe Biochimie des Produits Aquatiques (BPA), Boulevard du Bassin Napoléon, 62327 Boulogne-sur-Mer, France

[2] Université de Lorraine, UMR CNRS 7360 : Laboratoire Interdisciplinaire des Environnements Continentaux (LIEC), Rue Delestraint, 57070 Metz, France

Abstract: Glutathione S-transferases (GST), glutathione peroxidases (GPx) and metallothioneins (MT) are essential components of cellular detoxication systems. We studied the expression of pi-GST, Se-GPx, and MT transcripts in the digestive gland of *Dreissena polymorpha* exposed to organic and metallic pollutants. Mussels from a control site were transplanted during 3, 15 and 30 days into the Moselle River, upstream and downstream to the confluence with the Fensch River, a tributary highly polluted by polycyclic aromatic hydrocarbons and heavy metals. Se-GPx and pi-GST mRNA expression increased in mussels transplanted into the upstream site, Se-GPx response being the earliest. These genes were also induced after 3-days exposure at the downstream site. These inductions suggest an adaptative response to an alteration of the environment. Moreover, at this site, a significant decrease of the expression of MT, pi-GST and Se-GPx transcripts was observed after 30 days which could correspond to an inefficiency of detoxification mecanisms. The results are in correlation with the levels of pollutants in the sediments and their bioaccumulation in mussels, they confirm the environmental deleterious impact of the pollutants carried by the Fensch River.

Keywords: *Dreissena polymorpha*; pi-class GST; selenium-dependent GPx; metallothionein; mRNA levels; field study

---

1. Introduction

Industrial discharges are major sources of pollution for the freshwater aquatic environment.

Mixtures of pollutants produce numerous effects on organisms which can not be predicted exclusively with chemical analyses. To study the impact of chemical pollution the use of different biomarkers has been introduced into monitoring programs, in addition to chemical analyses, in order to evaluate the effects of pollutants on living species. To face pollutant effects, organisms possess cellular detoxication systems like glutathione S-transferases (GST), glutathione peroxidases (GPx) and metallothioneins (MT).

Bivalve molluscs are appropriate species to study the quality of the aquatic environment because they are sedentary and live in sediment or at the interface of water and sediment. They filter large amounts of water to cope with nutritional and respiratory needs. The zebra mussel *Dreissena polymorpha*, due to its abundance, distribution and functional role in ecosystems, had been often used to describe toxic effects in freshwater ecosystems. They are useful bioindicators to study the quality of an aquatic environment using active monitoring: mussels are sampled from a pristine area and caged in different sites for severals durations [1-6]. As transplanted mussels come from an homogeneous population, this approach allows to avoid the biological variability of organisms from various origins.

GST are cytosolic enzymes belonging to the most important phase II biotransformation system. They catalyse the conjugation of electrophilic compounds to glutathione. Among the enzymatic antioxidant system, the glutathione peroxidases can be divided into two types of enzymes: the selenium-dependent GPx (Se-GPx) and the selenium-independent GPx (non Se-GPx). Se-GPx catalyses the reduction of organic and inorganic peroxides like hydrogen peroxide (H_2O_2) while non Se-GPx reduces only organic peroxide [7].

Metallothioneins are low molecular weight proteins with high cystein content. The thiol groups (SH) of cystein residues enable MT to bind heavy metals. They are highly conserved and ubiquitously distributed throughout all organisms. They play a role in homeostatic control of essential metals (Cu, Zn) as they represent metal storage entities ready to fulfill enzymatic and metabolic demands [8]. MT are involved in metal detoxification and also protection against reactive oxygen species [9]. Monomeric (MT-10) and dimeric (MT-20) MT isoforms have been identified in mussels [10]. It was suggested that MT-10 takes part in the homeostasis of heavy metal cations, MT-20 being implicated in the defense against the effects of metals. Assessing pollution using MT is of great interest in the aquatic environment; indeed, invertebrate MT have been valided as useful biomarkers of exposure to metal contamination in molluscs and other aquatic organisms [11].

In order to evaluate the effect of industrial pollution on the detoxification performance of sentinel organisms, we studied the expression of pi-GST, Se-GPx and MT transcripts in the digestive gland of *D. polymorpha* transferred from a control site into the Moselle River, upstream and downstream from the confluence with the Fensch River. This river is located in a former industrial area in the Lorraine Region (France) and is polluted by high amounts of polycyclic aromatic hydrocarbons (PAHs) and heavy metals [12]. MT, pi-GST and Se-GPx coding sequences from the freshwater mussel *Dreissena polymorpha* have been previously identified [13,14]. These genes correspond to biomarkers widely used to evaluate exposure of aquatic organisms to numerous pollutants in the environment [15,16].

2. Materials and Methods

2.1. Experimental animals

Mussels, *D. polymorpha,* were sampled from a reference site, a small canal in the Meuse Basin (France). They were transplanted into the Moselle River upstream and downstream from the pollution source constituted by the confluence with the Fensch River (Figure 1). Mussels were kept in cages during 3, 15 or 30 days. Animals from the reference site served as controls. The digestive gland of control and transplanted mussels were removed immediately after collection, placed in 4 M guanidium isothiocyanate solution (Fermentas Life sciences, Vilnius, Lithuania) and conserved in liquid nitrogen to avoid RNA degradation. Five mussels were used at each site to measure the pi-GST, Se-GPx and MT mRNA expressions.

Figure 1. Location of the upstream and downstream sites in the Moselle River.

2.2. Chemical analysis

Analysis were performed on mussels soft tissues from control, upstream and downstream sites at each time of exposure, and on sediments samples collected at day 30 from upstream and downstream sites. Heavy metals were quantified by coupled plasma mass spectrometry (ICP-MS), atomic emission spectrometry (ICP-AES) or atomic absorption spectrometry (AAS). Polycyclic aromatic hydrocarbons were characterized by dichloromethane extraction followed by liquid chromatography and analysed by gas chromatography-mass spectrometry. The detection limits of these methods are indicated in Table 1.

Table 1. PAHs and metals concentrations in the sediments of the sites where mussels were encaged (dw: dry weight). The thresholds of average quality sediments correspond to the system established by the French water authorities (Agences de l'Eau, 2003). N.A.: no data available.

Contaminants	Concentration(mg/kg dw)		Quantification limits	Thresholds of average quality sediments (mg/kg dw)
	Upstream	Downstream		
PAHs			(µg/kg dw)	
Naphtalene	0.10	0.80	0.95	
Acenaphthylene	0.52	1.84	0.17	
Acenaphtene	0.14	0.70	0.22	
Fluorene	0.45	1.27	0.12	
Phenanthrene	3.42	5.77	0.17	
Anthracene	1.16	2.61	0.15	
Fluoranthene	7.79	17.51	0.31	
Pyrene	4.76	11.41	0.49	
Benzo(a)anthracene	3.03	7.49	0.18	
Chrysene	2.78	6.98	0.24	
Indeno(1,2,3-cd)pyrene	1.94	5.80	30.00	
Benzo(b)fluoranthene	3.07	8.43	26.00	
Benzo(k)fluoranthene	1.16	3.30	28.00	
Benzo(a)pyrene	1.45	5.37	25.00	
Dibenzo(a,h)anthracene	0.75	2.20	81.00	
Benzo(g,h,i)perylene	1.55	4.67	0.34	
Total PAHs	34.07	86.15		7.5
Metals			(mg/kg dw)	
Chromium	111	758	5	110
Cooper	39	135	1	140
Lead	137	428	0.6	120
Nickel	37	103	5	48
Tin	10	741	0.5	N.A.
Zinc	498	2138	1	460

2.3. RNA isolation

RNA extraction was carried out on 30 mg of digestive gland using the RNeasy Mini Kit (Qiagen, Maryland, USA) following the manufactor instructions. An electrophoresis on agarose gel was carried out in order to check the integrity of the RNAs.

2.4. Primers design

Primers were obtained from Invitrogen (Carlsbad, USA), the sequences are given in Table 2.

They were choosen in the beta-actin, MT, pi-GST and Se-GPx coding sequences available in GenBank or identified in our laboratory (accession no: AF082863, DPU67347, EF194203, EF194204 respectively).

Table 2. Sequences of primers used for amplification of *D. polymorpha* cDNA.

Primers	Sequences (5'-3')
ACT.DF	GGATCTGGAATGTGCAAAG
ACT.DR	CATCCCAGTTGGTGACGATA
MT.DF	ACCTGACTTTACGCATTCAAC
MT.DR	AACTGTGTGACACCATCCCA
GST.DF	TG ACTTGATCAAGGACGCGA
GST.DR	GATAGCACGCACAGGATGTC
GPX.DF	GGAGTTGACGAACGGTCTA
GPX.DR	GATGGGATGGTACAGCTTCT

2.5. cDNA synthesis

Reverse-transcriptions were performed on 1 μg of total RNA using the Maxime RT oligo dT PreMix Kit (iNtRON Biotechnology, Seongnam, South Korea) according to the manufacturer's instructions. At the end of the reaction each cDNA were diluted 10 times in Tris HCl 10 mM (pH 8.3).

2.6. MT, pi-GST and Se-GPx mRNA expression

Four μL of cDNA were diluted from 1:2 to 1:64 to calculate PCR efficiencies which are the following: 1.081 (MT), 1.029 (pi-GST), 0.928 (Se-GPx) and 1.030 (beta-actin).

Amplification of MT, pi-GST, Se-GPx and beta-actin cDNA was carried out by real-time quantitative PCR in a MiniOpticon System (Bio-Rad, Hercules, USA) using GoTaq qPCR Master Mix containing SYBR Green (Promega, Madison, USA) according to the manufacturer's instructions. PCR amplifications were carried out on 4 μL of 8-folds diluted cDNA using 300 nM of beta-actin or MT primers, or 500 nM of pi-GST or Se-GPx primers. The PCR steps consisted of a 10-min initial denaturation at 95 °C followed by 40 cycles of heat denaturation at 95 °C for 30 s, annealing at 58 °C for 30 s, polymerisation at 72 °C for 30 s, and a 5-min final extension at 72 °C. Negative controls were performed on non reverse-transcribed RNA.

The normalized folds of MT, pi-GST and Se-GPx transcripts were calculated using the -ΔΔCt method [17].

2.7. Statistical analyses

Statistical analyses were carried out to compare MT, pi-GST and Se-GPx mRNA expression in control and transplanted mussels. The ratios of MT, pi-GST and Se-GPx levels/beta-actin levels were analysed by a t-test (comparison of means) using the R software [18]. As PCR is a geometric reaction, we considered differences as significant if the fold change was ≥ 1.5 and $P \leq 0.05$.

3. Results

3.1. Chemical analysis

The concentrations of pollutants in sediments from upstream and downstream sites are given in Table 1. Total PAHs concentration in the upstream and downstream sites are respectively around 34 and 86 mg/kg of dry sediment. These sites are also contaminated by heavy metals. For example, chromium, lead and zinc concentrations in the downstream site are respectively 6, 3 and 4 times higher than the thresholds of average quality sediments.

The analysis show a bioaccumulation of metals and PAHs in mussels in accordance with the time of exposure (Table 3). The highest levels of almost all metals and PAHs are pointed out in mussels exposed 30 days at the downstream site. For instance, total PAHs, copper and zinc concentrations are 2-times the values in control mussels whereas iron and lead levels are around 10-times higher compared to controls.

Table 3. Concentrations of total PAHs and metals in the soft tissues of control and transplanted mussels (dw: dry weight).

Pollutants	Concentrations in mussels (µg/g dw)						
	Control site	Upstream site			Downstream site		
		3 days	15 days	30 days	3 days	15 days	30 days
Total PAHs	32.93	32.99	50.50	45.60	35.27	67.43	65.22
Cd	0.40	0.33	0.31	0.66	0.40	0.48	0.40
Cu	8.97	9.14	16.68	17.82	12.06	16.99	20.32
Fe	118.30	227.91	374.89	584.96	410.10	638.38	1089.76
Ni	4.93	4.44	4.92	7.54	5.12	5.17	6.27
Pb	0.23	0.34	0.52	1.07	0.82	1.32	2.20
Zn	83.25	88.97	114.45	185.37	103.88	166.04	169.48

3.2. MT, pi-GST and Se-GPx expression patterns

We evaluated the expression pattern of MT, pi-GST and Se-GPx transcripts from *D. polymorpha* in the upstream, downstream sites (Figure 2). Amplifications performed on non reverse-transcribed RNA (negative control) did not produce any amplimer, indicating that PCR products were not amplified from any genomic DNA remaining in the samples.

Mussels transplanted into the upstream site showed an increase of pi-GST mRNA levels after 30 days of exposure, and as soon as 3 days in the case of the Se-GPx compare to the controls. Both genes were induced after 3-days exposure at the downstream site. A significant decrease of MT, pi-GST and Se-GPx mRNA expression levels was observed at the downstream site after 30 days.

Considering the amounts of pollutants in *D. Polymorpha* soft tissues, no clear correlation was established between genes expression and PAHs concentrations. However, a binomial curve was observed in the case of metals (Figure 3), correlation coefficients of MT, pi-GST and Se-GPx expressions beeing respectively 0.921, 0.726 and 0.997.

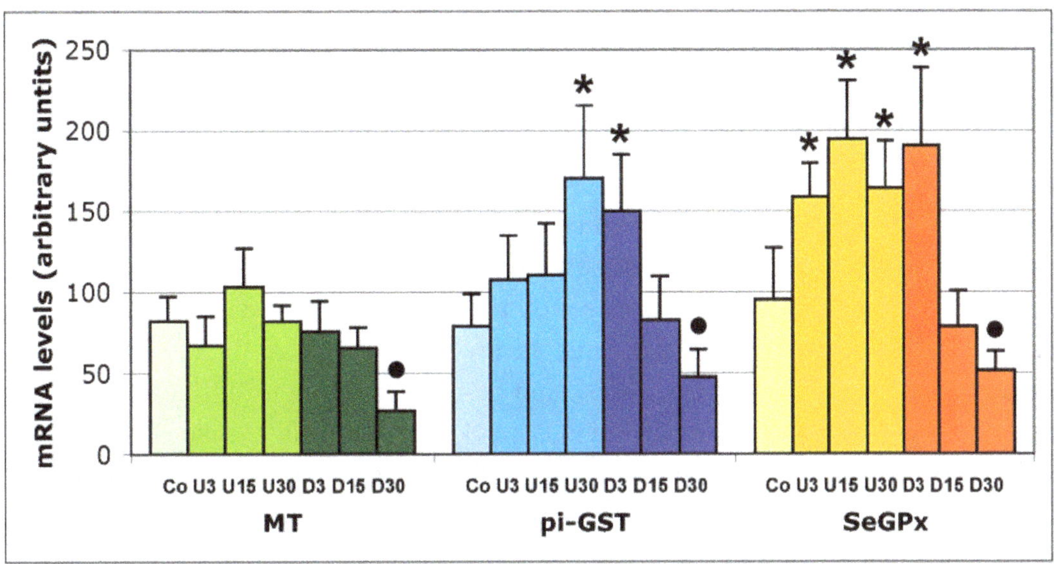

Figure 2. Expression levels of MT, pi-GST and Se-GPx mRNA in the digestive gland of *D. polymorpha* from control site (Co) and transplanted 3, 15 or 30 days into the Moselle River. U and D correspond to upstream and downstream sites. Stars (*) and dots (•) indicate respectively significant increase or decrease compared to control.

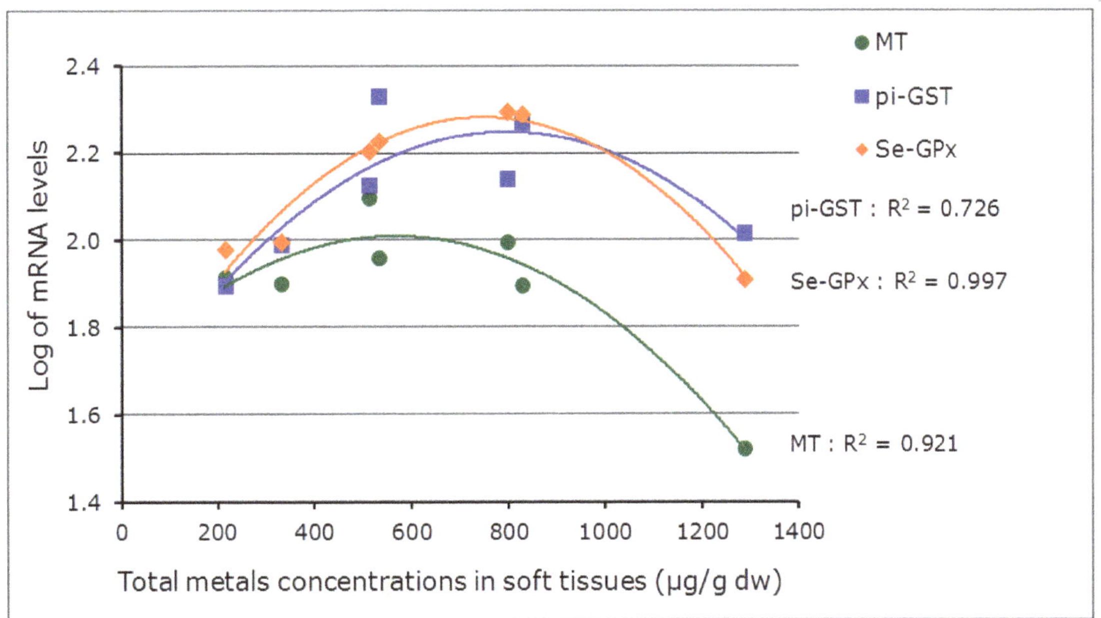

Figure 3. MT, pi-GST and Se-GPx mRNA expressions according to total metals concentrations in mussels soft tissues.

4. Discussion

The Fensch River is the major source of pollution in the present field study. The bed and the banks of this river are highly degraded by urbanisation and industrialisation [19]. A large part of the pollutants is rapidly transferred into the Moselle River. The pollutant concentrations in downstream site (Table 1) largely exceed the thresholds corresponding to sediments of average quality according to the river quality classification system established by the French Water Authorities [20].

MT is a scavenger of heavy metals and free radicals, Se-GPx is a major component of the antioxidant systems and pi-GST is a phase II biotransformation enzyme. As these proteins are involved in different detoxification process, their expression could be used as warning indicator of environmental contamination.

The significant increase of pi-GST and Se-GPx mRNA levels in mussels transplanted into upstream and downstream sites could correspond to an antioxidant stress response. However, MT, pi-GST and Se-GPx expression showed a decrease trend at the downstream site according to the time of exposure. We observed a similar hormesis effect on mussels, *C. fluminea*, exposed to copper and cadmium [21]. The transcript levels of the three genes were significantly lower after 30-days exposure compared to controls. Moreover, pi-GST and Se-GPx expressions at days 15 and 30 were lower compared to day 3. The results of a previous work performed in the Fensch River by our laboratory are in correlation with the present study: freshwater mussels, *U. tumidus*, encaged downstream from the outfall of a cocking plant exhibited lower Se-GPx activities in the digestive gland compared to values at the upstream site [22].

It has been demonstrated that GST and GPx activities of mussels were induced by environmental pollutants [23-28] and that MT synthesis was induced by metal contaminants [29]. However, other studies showed also a decrease of GST, GPx or MT at the protein or mRNA levels. This decrease of MT, pi-GST and Se-GPx transcripts levels in mussel transplanted into the downstream site could correspond to an overwhelming of the detoxification systems; il could also be due to the presence of inhibitors of their expression in higher proportions than activators in the pollutant mixture. Indeed, pi-GST gene expression is under the regulatory control of both enhancer and silencer elements [30].

A decrease of GST activity was observed in a freshwater mussel, *Anodonta cygnea*, exposed to pesticides [31]. Results similar to our study were obtained with the mollusc *Ruditapes decussatus* exposed to three concentrations of municipal effluents for 7 and 14 days: GST activity was induced at the lowest concentration and time-exposure, then decreased in animals exposed to the two other concentrations for 14 days [32]. Such biphasic response was pointed out in the digestive gland of snails exposed to several levels of metals [33].

An exposure to copper, performed in laboratory, showed a significant decrease of MT protein level in limpets *Patella vulgata* [34]. Mussels, *Mytilus galloprovincialis*, exposed to environmentally relevant concentrations of chromium showed a decrease of MT mRNA level [35]. An inhibition of MT expression was pointed out in the oyster *Crassostrea gigas* from an area higly polluted by metals compared to control organisms [36].

Several authors have shown a decrease of GPx activity in organisms exposed to multiple water pollutants. For example, the bivalves *Ruditapes decussatus* collected from sites contaminated by metals and PAHs showed a lower GPx activity compared to control organisms [37]. Heavy metals can also lead to a decrease of GPx activity. Mussels, *Mytilus galloprovincialis*, from the Izmir Bay revealed a decrase of GPx activity in animals sampled in an heavily polluted area compared to a clean area [38]. These observations showed the relationship between the proximity of the pollution source and the responses of the organisms.

High toxicant concentrations may lead to the inhibition of enzyme activity [39]. For example, a reduced GPx activity could indicate that its antioxidant capacity was surpassed by the amount of hydroperoxides produced by lipid peroxidation [40]. Moreover, a deficiency of the Se-GPx activity reduces detoxification efficiency and reveals a precarious status in the exposed bivalves, suggesting that this inhibition was predictive of toxicity [22]. A reduced capability to neutralise radical oxidative species has been suggested to play a fundamental role in mediating oxidative toxicity and has been often associated with the presence of cellular alterations [41].

The present work showed the deleterious effects of an industrial pollution on the detoxification capacity of aquatic organisms. The results are in accordance with the amounts of PAHs and metals identified in the sediments and with the bioaccumulation of pollutants in transplanted mussels. Detoxification systems were induced in mussels transplanted into the upstream site wich showed moderate levels of pollutants. However, after an early increase, an inhibition of these systems was pointed out in dreissena encaged 30 days in the highly polluted downstream site. These results were likely to correspond to an overwhelming of MT, pi-GST and Se-GPx defense capabilities suggesting that mussels were not able to face the environmental perturbations. A biphasic response was observed according to metals concentration in mussels soft tissues, especially in the cases of MT and Se-GPx (Figure 3), suggesting that, in the present study, metallics pollutants have more impact on dreissena detoxification systems than PAHs. Therefore, it is also highly likely that transplanted mussels are mainly exposed to an oxidative stress. Indeed, previous studies performed at the same sites showed lipid peroxidation and depletion of antioxidant defense systems in transplanted mussels [42].

Our results showed that molecular methods, such as study of detoxification genes expression in sentinel organisms, could be an early warning of environmental perturbations. In the perspective of the present study, an other downstream site more distant to the confluence with the Fensch River could be chosen to assess if the modulation of gene expression is linked to the distance with the source of pollution.

5. Conclusion

We used *D. Polymorpha* to monitor the effects of an industrial pollution on the detoxification performance of sentinel organisms. Mussels were transplanted into the Moselle River upstream and downstream from the confluence with the Fensch River, a tributary polluted by PAHs and heavy metals. Modulation of MT, pi-GST and Se-GPx mRNA levels correlates with the amounts of pollutants (especially metals) and the exposure duration. These results confirm the environmental deleterious impact of the pollutants carried by the Fensch River.

The present study shows that molecular approaches performed on bioindicators such as *D. polymorpha* are essential tools to detect environmental perturbations. Expression studies of pi-GST, Se-GPx and MT at the transcriptional level will also contribute to a better insight of the xenobiotic toxicity in the aquatic environment.

Acknowledgements

This work was supported by the CPER of the Region Lorraine and the French Ministry of Research.

Conflict of interest

All authors declare no conflicts of interest in this paper.

References

1. Roméo M, Hoarau P, Garello G, et al. (2003) Mussel transplantation and biomarkers as useful tools for assessing water quality in the NW Mediterranean. *Environ Pollut* 122: 369-378.
2. Guerlet E, Ledy K, Meyer A, et al. (2007) Towards a validation of a cellular biomarker suite in native and transplanted zebra mussels: A 2-year integrative field study of seasonal and pollution-induced variations. *Aquat Toxicol* 81: 377-388.
3. Viarengo A, Lowe D, Bolognesi C, et al. (2007) The use of biomarkers in biomonitoring: A 2-tier approach assessing the level of pollutant-induced stress syndrome in sentinel organisms. *Comp Biochem Physiol C Toxicol Pharmacol* 146: 281-300.
4. Borcherding J (2010) Steps from ecological and ecotoxicological research to the monitoring for water quality using the zebra mussel in a biological warning system, In: Van der Velde G, Rajagopal S and bij de Vaate, *A The Zebra Mussel in Europe*, Leiden/Margraf, Weikersheim: Backhuys, 279-283.
5. Voets J, Bervoets L, Smolders R, et al. (2010) Biomonitoring environmental pollution in freshwater ecosystems using Dreissena polymorpha, In: Van der Velde G, Rajagopal S and bij de Vaate, *A The Zebra Mussel in Europe*, Leiden/Margraf, Weikersheim: Backhuys, 301-321.
6. Chatel A, Faucet-Marquis V, Gourlay-Francé C, et al. (2015) Genotoxicity and activation of cellular defenses in transplanted zebra mussels Dreissena polymorpha along the Seine River. *Ecotoxicol Environ Saf* 114: 241-249.
7. Almar M, Otero L, Santos C, et al. (1998) Liver glutathione content and glutathione-dependent enzymes of two species of freshwater fish as bioindicators of chemical pollution. *J Environ Sci Health* 33: 769-783.
8. Viarengo A, Nott JA (1993) Mechanisms of heavy metal cation homeostasis in marine invertebrates. *Comp Biochem Physiol C Toxicol Pharmacol* 104: 355-372.
9. Roesijadi D (1996) Metallothionein and its role in toxic metal regulation. *Comp Biochem Physiol C Toxicol Pharmacol* 113: 117-123.
10. Lemoine S, Laulier M (2003) Potential use of the levels of the mRNA of a specific metallothionein isoform (MT-20) in mussel (*Mytilus edulis*) as a biomarker of cadmium contamination. *Mar Pollut Bull* 46: 1450-1455.
11. Mourgaud Y, Martinez E, Geffard A, et al. (2002) Metallothionein concentration in the mussel *Mytilus galloprovincialis* as a biomarker of response to metal contamination: validation in the field. *Biomarkers* 7: 479-490.
12. Rodius F, Hammer C, Vasseur P (2002). Use of RNA arbitrarily primed PCR to identify genomic alterations in the digestive gland of the freshwater bivalve *Unio tumidus* at a

contaminated site. *Environ Toxicol* 17: 538-546.

13. Engelken J, Hildebrandt A (1999) cDNA cloning and cadmium-induced expression of metallothionein mRNA in the zebra mussel *Dreissena polymorpha. Biochem Cell Biol* 77: 237-241.
14. Doyen P, Bigot A, Vasseur P, et al. (2008) Molecular cloning and expression study of pi-class glutathione S-transferase (pi-GST) and selenium-dependent glutathione peroxidase (Se-GPx) transcripts in the freshwater bivalve *Dreissena polymorpha. Comp Biochemi Physiol C Toxicol Pharmacol* 147: 69-77.
15. Snell TW, Brogdon SE, Morgan MB (2003) Gene expression profiling in ecotoxicology. *Ecotoxicology* 12: 475-483.
16. Sarkar A, Ray D, Shrivastava AN, et al. (2006) Molecular Biomarkers: their significance and application in marine pollution monitoring. *Ecotoxicology* 15: 333-340.
17. Livak KJ, Schmittgen TD (2001) Analysis of relative gene expression data using real-time quantitative PCR and the $2^{-\Delta\Delta Ct}$ method. *Methods* 25: 402-8.
18. R, Developement Core Team (2006) R: A 701 language and environment for statistical 702 computing. R foundation for Statistical Computing, Vienna, Austria. ISBN 3-900051-07-0
19. Mazauer P, Matte JL (2001) Qualité du milieu physique de la Fensch – synthèse. Direction Régionale de l'Environnement de Lorraine, France.
20. Agences de l'Eau (2003) Seuils de qualité pour les micropolluants organiques et minéraux dans les eaux superficielles-synthèse, In: Agences de l'Eau, *Système d'Evaluation de la Qualité de l'Eau*, Paris, France.
21. Bigot A, Minguez L, Giamberini L, et al. (2011) Early defense responses in the freshwater bivalve Corbicula fluminea exposed to copper and cadmium: transcriptional and histochemical studies. *Environ Toxicol* 26: 623-632.
22. Cossu C, Doyotte A, Jacquin MC, et al. (1997) Glutathione reductase, selenium-dependent glutathione peroxidase, glutathione levels, and lipid peroxidation in freshwater bivalves, *Unio tumidus*, as biomarkers of aquatic contamination in field studies. *Ecotoxicol Environ Saf* 38: 122-131.
23. Suteau P, Daubeze M, Migaud ML, et al. (1988) PAH metabolising enzymes in whole mussel as biochemical tests for chemical pollution monitoring. *Mar Ecol Prog Ser* 46: 45-49.
24. Livingstone DR, Martinez PG, Michel X, et al. (1990) Oxyradical production as a pollution-mediated mechanism of toxicity in the common mussel, *Mytilus edulis L.*, and other molluscs. *Funct Ecol* 4: 415-424.
25. Livingstone DR (1998) The fate of organic xenobiotics in aquatic ecosystems: quantitative and qualitative differences in biotransformation by invertebrates and fish. *Comp Biochem Physiol A Mol Integr Physiol* 120: 43-49.
26. Sole M, Porte C, Albaiges J (1994) Mixed-function oxygenase system components and antioxidant enzymes in different marine bivalves: its relation with contaminants body burdens. *Aquat Toxicol* 30: 271-283.
27. Chatel A, Faucet-Marquis V, Perret M, et al. (2012) Genotoxicity assessement and detoxification induction in Dreissena polymorpha exposed to benzo[a]pyrene. *Mutagenesis* 27: 703-711.
28. Chatel A, Faucet-Marquis V, Pfohl-Leszkowicz A, et al. (2014) DNA adduct formation and induction of detoxification mechanisms in Dreissena polymorpha exposed to nitro-PAHs. *Mutagenesis* 29: 457-465.

29. Bebianno MJ, Langston WJ (1992) Cadmium induction of metallothionein synthesis in *Mytilus galloprovincialis*. *Comp Biochem Physiol C Toxicol Pharmacol* 103: 79-85.
30. Imagawa M, Osada S, Koyama Y, et al. (1991) SF-B that binds to a negative element in glutathions transferase P gene is similar or identical to trans-activator LAP/IL6-DBP. *Biochem Biophys Res Commun* 179: 293-300.
31. Robillard S, Beauchamp G, Laulier M (2003) The role of abiotic factors and pesticide levels on enzymatic activity in the freshwater mussel *Anodonta cygnea* at three different exposure sites. *Comp Biochem Physiol C Toxicol Pharmacol* 135: 49-59.
32. Kamel N, Jebali J, Banni M, et al. (2012) Biochemical responses and metals levels in Ruditapes decussatus after exposure to treated municipal effluents. *Ecotox Env Saf* 82: 40-46.
33. Radwan MA, El-Gendy KS, Gad AF (2010) Oxidative stress biomarkers in the digestive gland of Theba pisana exposed to heavy metals. *Arch Environ Contam Toxicol* 58: 828-835.
34. Brown RJ, Galloway TS, Lowe D, et al. (2004) Differential sensitivity of three marine invertebrates to copper assessed using multiple biomarkers. *Aquat Toxicol* 66: 267-278.
35. Ciacci C, Barmo C, Gallo G, et al. (2012) Effects of sublethal, environmentally relevant concentrations of hexavalent chromium in the gills of Mytilus galloprovincialis. *Aquat Toxicol* 120: 109-118.
36. David E, Tanguy A, Riso R, et al. (2012) Responses of Pacific oyster Crassostrea gigas populations to abiotic stress in environmentally contrasted estuaries along the Atlantic coast of France. *Aquat Toxicol* 109: 70-79.
37. Geret F, Serafim A, Bebianno J (2003) Antioxidant enzyme activities metallothioneins and lipid peroxidation as biomarkers in *Ruditapes decussatus*? *Ecotoxicology* 12: 417-426.
38. Kucuksezgin F, Kayatekin BM, Uluturhan E, et al. (2008) Preliminary investigation of sensitive biomarkers of trace metal pollution in mussel (*Mytilus galloprovincialis*) from Izmir Bay (Turkey). *Environ Monit Assess* 141: 339-345.
39. Cheung CCC, Zheng GJ, Li AM, et al. (2001) Relationships between tissue concentrations of polycyclic aromatic hydrocarbons and antioxidative responses of marine mussels, Perna viridis. *Aquat Toxicol* 52: 189-203.
40. Remacle J, Lambert D, Raes M, et al. (1992) Importance of various antioxidant enzymes for cell stability. Confrontation between theoretical and experimental data. *Biochem J* 286: 41-46.
41. Regoli F (2000) Total oxyradical scavenging capacity (TOSC) in polluted and translocated mussels: a predictive biomarker of oxidative stress. *Aquat Toxicol* 50: 351-61.
42. Doyotte A, Cossu C, Jacquin MC, et al. (1997) Antioxidant enzymes, glutathione and lipid peroxidation as relevant biomarkers of experimental or field exposure in the gills and the digestive gland of the freshwater bivalve Unio tumidus. *Aquatic Toxicol* 39: 93-110.

10

Ethanol production from hot-water sugar maple wood extract hydrolyzate: fermentation media optimization for *Escherichia coli* FBWHR

Yang Wang, Chenhui Liang and Shijie Liu

Department of Paper and Bioprocess Engineering, State University of New York—College of Environmental Science and Forestry, Syracuse, NY 13210, USA

Abstract: We report the first time statistical study of the optimization for ethanol production from hot-water sugar maple hemicellulosic wood hydrolyzate by *Escherichia coli* FBWHR. Response surface methodology was employed to investigate the effect of fermentation media on the ethanol production from concentrated hot-water sugar maple hemicellulosic wood extract hydrolyzate by *Escherichia coli* FBWHR. The critical media components were firstly selected according to Plackett–Burman design and further optimized by central composite design. Based on the response surface analysis, the optimum concentrations of the significant components were obtained: yeast extract, 10.19 g/L; tryptone, 14.55 g/L; $Na_2HPO_4 \cdot 7H_2O$, 23.21 g/L; KH_2PO_4, 5 g/L and NH_4Cl, 2 g/L. An ethanol concentration of 15.23 ± 0.21 g/L was achieved under the optimized media, which agreed with the predicted value. Ethanol production was enhanced to 22.18 ± 0.13 g/L by scaling up the fermentation from shaker flask to 1.3 L bioreactor.

Keywords: hot-water sugar maple wood hydrolyzate; recombinant *Escherichia coli*; ethanol fermentation; response surface methodology

---

1. Introduction

Development of bio-fuels has recently drawn attention due to the depletion of crude oil, increased oil prices and environmental benefits [1-4]. For example, ethanol as a renewable energy source as well as an important industrial ingredient has been produced from different kinds of raw materials such as urban waste, agricultural residue and forest materials which are counted low-cost lignocellulosic biomass [5,6]. Woody biomass, comprising about 40% hemicellulose with the major components of hexoses (glucose, galactose, mannose, rhamnose) and pentoses (xylose and arabinose), is the most abundant organic source on Earth and has been widely studied in bio-ethanol production process [7-10]. Recently, our group has reported an ethanol batch fermentation using a new improved strain adapted from FBHW, *E. coli* FBWHR, from hot-water sugar maple wood extract hydrolyzate and its kinetic study [11].

The optimization of fermentation media is very important to the ethanol process by possibly improving the consumption of the woody biomass constituents and maximizing the ethanol production. The classical method of "one factor at a time" is laborious and time-consuming, especially when the number of variables is large. Design of experiments (DoE) is a structured and robust methodology for optimizing fermentation processes and media and well documented in the literature [12,13,14]. Plackett–Burman (PB) design, involving a two-level fractional factorial saturated design, tests the largest number of factor effects with the least number of observations and screens for significant factors [15]. It has been proved to be useful in evaluating the relative significance of variables and screening a large number of media components [16,17]. Response surface methodology (RSM) is followed for identifying the effect of individual variables and for seeking the optimal fermentation conditions for a multivariable system with minimum numbers of experiments efficiently. It has been successfully adopted to optimize the production of ethanol by different microorganisms [18-21]. Central composite design (CCD) is the most widely used response surface design that provides statistical models which help in understanding the interactions among the parameters that have been optimized [22,23]. RSM uses CCD to fit a model by the least squares technique. Three major steps are involved in RSM: performing the statistically designed experiments, estimating the coefficients in a mathematical model, and predicting the response and checking the adequacy of the model.

However, there are no literature reports on media optimization of ethanol fermentation from hot-water sugar maple hemicellulosic wood extract hydrolyzate. In this study, we report the optimization of the fermentation media specifically comprising hot-water sugar maple hemicellulosic wood extract hydrolyzate and a new improved strain adapted from FBHW, *E. coli* FBWHR by statistical design of experiments. To enhance the production of ethanol using RSM, a PB design was performed to screen for components of the production medium that had significant effects on the ethanol production. A CCD was then employed to optimize the significant factors identified by PB design to further increase ethanol production. Finally, batch fermentation was scaled up in a 1.3 L fermenter.

2. Material and Methods

2.1. Sugar maple wood extract hydrolyzate, E. coli FBWHR strain and cell growth

The details of preparation of sugar maple wood extract hydrolyzate, microorganism and cell growth, strain adaptation, and analytical methods have been reported by Wang et al. [11]. The hot-water extraction was carried out in a 1.84 m^3 digester with a wood to liquor ratio of 1:4 at about 160 °C for 2 h, followed by Ultra-filtration and Nano-filtration. The concentrated extracts were hydrolyzed at 135 °C for 25 min with 1 wt% sulfuric acid added to remove lignin. Hydrolyzate was neutralized by $Ca(OH)_2$ and then fractionated by Nano-filtration. According to previous investigation, hot-water sugar maple hemicellulosic wood extracts hydrolyzate used in this study contained six monosaccharides, primarily xylose and minor amounts of glucose, mannose, arabinose, galactose and rhamnose, other compounds such as phenolics and aromatics, and several trace metals [11]. *E. coli* FBWHR was grown in 250 mL shaker flasks containing 100 mL sterile liquid medium (10 g/L tryptone, 5 g/L yeast extract, and 5 g/L NaCl) adjusted to a pH of 7.0, and supplemented with 10 g/L xylose and 20 mg/L tetracycline. The liquid culture was incubated at 35 °C for 12–14 hours on an incubator shaker with a speed of 160 rpm. After growth, 5 mL of culture was used to inoculate 100 mL of sterile fermentation media.

2.2. Ethanol fermentation conditions

2.2.1. Ethanol fermentation in shaker flasks

Concentrated hot-water sugar maple wood extract hydrolyzate was diluted to 30% (v/v) with different concentrations of defined media solutions (100 mL), and then transferred into a 250 mL shake flask. The pH of the fermentation media was adjusted to 7.0 by addition of ammonia hydroxide. Concentrated hot-water sugar maple wood extract hydrolyzate and ammonia hydroxide were filter-sterilized. Other media solutions were autoclaved separately at 121 °C for 15 min. Fermentations studied by RSM were conducted at 35 °C with the incubator shaker at a shaking speed of 160 rpm.

2.2.2. Ethanol fermentation in bioreactor

Batch fermentation was scaled up from shaker flasks to a 1.3 L New Brunswick Bioreactor (BIOFLO 110) with a working volume of 800 mL under micro-aerobic condition. Sugar concentration of fermentation media was achieved by diluting concentrated hydrolyzate to 30% (v/v) containing 48.53 g/L total sugar. After cooling, concentrated hydrolyzate was added to the bioreactor and adjusted to pH 7.0 by ammonia hydroxide prior to inoculation. Samples (2 mL) were taken intermittently and centrifuged prior to analysis.

2.3. Analytical methods

Cell density (g/L) was estimated by using a predetermined correlation between dry weight cell concentrations (oven dry at 105 °C) versus optical density. Ethanol concentration was measured by GC using Thermo Scientific Focus GC systems equipped with a Triplus automatic sampler and a TRACE TR-WaxMS (30 m × 0.25 mm × 0.25 μm) GC column. Sugar concentrations were determined by nuclear magnetic resonance (NMR) spectroscopy using a modified two dimensional heteronuclear single quantum coherence (HSQC) experiment.

Element concentrations of the concentrated hot-water sugar maple wood extract hydrolyzate were analyzed by inductively coupled plasma (ICP) on a Perkin Elmer 3300DV Inductively Coupled Plasma Emission Spectrometer. Samples were diluted 100-fold with 2% nitric acid for analysis. The element concentrations were obtained via emissions.

2.4. Experimental design

2.4.1. Plackett–Burman (PB) design

Six medium components (yeast extract, tryptone, $Na_2HPO_4{\cdot}7H_2O$, KH_2PO_4, NH_4Cl and NaCl) were chosen in the present study for the effect on the ethanol production by hot-water sugar maple wood extract hydrolyzate. The main effect of variables was firstly determined as follows:

$$E_{(xi)} = \frac{(\sum y_{i+} - \sum y_{i-})}{N/2} \quad (1)$$

Where, $E_{(xi)}$ represents the effect of variable *i*; y_{i+} is the response of the high level of variable *i*; y_{i-} is the response of the low level of variable *i*; N stands for the total number of trials.

The standard error (SE) of the effect was the square root of an effect, and the significance level (*p*-value) of each variable effect was determined by *T*-test as follows:

$$t_{(xi)} = \frac{E\,(xi)}{SE} \quad (2)$$

Where, $E_{(xi)}$ represents the effect of variable *i*; SE is the standard error [17,24].

2.4.2. Central composite design (CCD)

After the critical medium components were screened by PB design, Response Surface Methodology (RSM) was adopted to optimize the concentrations of the components to maximize the production of ethanol. Three components (yeast extract, tryptone, and $Na_2HPO_4{\cdot}7H_2O$) that significantly affected the ethanol production were optimized by RSM using a 3-factor-5-level CCD with 6 replications of center points. The CCD matrix included six central points and six axial points, with an axial distance of ±1.68 to make the design orthogonal [25]. In order to develop the second-order regression model, the variables were coded to the following equation:

$$x_i = (X_i - X_0)/\Delta X_i,\ i = 1,2,\ldots,k \quad (3)$$

Where x_i and X_i are the dimensionless and the actual values of the independent variable *i*, X_0 is the actual value of the independent variable at the center point, and ΔX_i is the step change of X_i corresponding to a unit variation of the dimensionless value. In order to correlate the response variable (i.e., ethanol concnetration) to the independent variables, second-order polynomial model for predicting the optimal point was expressed as follows:

$$Y = b_0 + \sum b_i x_i + \sum b_{ii} x_i^2 + \sum b_{ij} x_i x_j,\ i = 1,2,\ldots,k;\ j = 1,2,\ldots,k,\ i \neq j \quad (4)$$

Where Y represents the predicted response, b_0 is the interception coefficient, b_i is the coefficient of the linear effect, b_{ii} is the coefficient of the quadratic effect, b_{ij} is the coefficient of the interaction effect when $i < j$, and k is the numbers of the involved variables.

2.5. Statistical analysis

The experiments were carried out in triplicate unless other specified. The average ethanol concentration was used as the dependent variable. The Design-Expert 8.0.7.1 trial software was used for multiple regression analysis of the experimental data. The analysis of variance (ANOVA) was evaluated. The statistical and regression coefficient significance was checked with F-test and t-test, respectively. The multiple coefficients of correlation (R) and the determination coefficient of correlation (R^2) were calculated to evaluate the adequacy of the model. The optimum values of the selected variables were obtained by analyzing the response surface plot and solving the regression equation.

3. Results and Discussion

3.1 Screening of significant media components that enhance ethanol production

PB design experiments were performed to select the important fermentation parameters which have significantly improved the ethanol production by *E. coli* FBWHR. An initial set of six media components with two levels of concentration was proposed with a potential effect on the ethanol fermentation. Yeast extract and tryptone are widely employed as complex sources of peptides, amino acids, vitamins, and trace elements such as magnesium for microorganisms in fermentation media [26,27]. It has been reported that cells have to synthesize amino acid and vitamin and require more energy and carbon source without the supplement of yeast extract [28]. Na_2HPO_4, KH_2PO_4 NH_4Cl and NaCl are the typical substances for *E. coli* growth. The experimental design matrix based on the Plackett–Burman design was presented in Table 1. Twelve experiments were carried out using different combinations of the variables. The results were analyzed by using the analysis of variance (ANOVA) as appropriate to the experimental design used. The sum of squares, mean squares, F-values and *p*-values were estimated. It was demonstrated that the model was highly significant with the F-value of 19.39 and a very low probability value of 0.0062 (p-value < 0.05) meaning only a 0.62% chance that a “Model F-Value” this large could occur due to noise. According to the ANOVA results (Table 2), yeast extract, tryptone and $Na_2HPO_4 \cdot 7H_2O$ had significant positive effects on the ethanol production, because the values of “Prob > F” were less than 0.05. KH_2PO_4 and NH_4Cl also had positive effects but the contributions were not significant, so their concentrations were fixed at the high level and entered to the CCD experiments. NaCl was the only media component that had an insignificantly negative impact and a low level of 0 g/L. Therefore, NaCl was removed from the media fomulation. The determination coefficient (R^2) was 0.9714, which meant that the model could explain 97.14% of the total variations in the system. The predicted determination coefficient (Pred $R^2 = 0.7277$) was in reasonable agreement with the adjusted determination coefficient (Adj $R^2 = 0.9213$). The signal to noise ratio was measured by “Adeq Precision”. A ratio greater than 4 is desirable. A ratio of 12.497 indicated an adequate signal,which implied that this model could be used to navigate the design space. The optimum values of yeast extract, tryptone and $Na_2HPO_4 \cdot 7H_2O$ were further investigated by CCD.

Table 1. The PB design for screening variables in ethanol production from 30% (v/v) of hot-water sugar maple hemicellulosic wood hydrolysate by *E. coli* FBWHR.

Run	Levels	Yeast Extract (X_1), g/L	Tryptone (X_2), g/L	$Na_2HPO_4 \cdot 7H_2O$ (X_3), g/L	KH_2PO_4 (X_4), g/L	NH_4Cl (X_5), g/L	NaCl (X_6), g/L	Ethanol concentration, g/L ($n = 3$)
	−1	1.00	1.00	1.00	1.00	0.50	0.00	
	+1	10.00	10.00	20.00	5.00	2.00	10.00	
1		1.00	10.00	1.00	5.00	2.00	0.00	1.56 ± 0.10
2		1.00	10.00	20.00	5.00	0.50	0.00	5.88 ± 0.24
3		1.00	1.00	20.00	1.00	2.00	10.00	1.31 ± 0.15
4		10.00	10.00	1.00	5.00	2.00	10.00	1.09 ± 0.12
5		1.00	10.00	20.00	1.00	2.00	10.00	6.79 ± 0.18
6		10.00	1.00	20.00	5.00	0.50	10.00	2.67 ± 0.08
7		1.00	1.00	1.00	1.00	0.50	0.00	0.28 ± 0.10
8		10.00	1.00	1.00	1.00	2.00	0.00	3.74 ± 0.05
9		10.00	1.00	20.00	5.00	2.00	0.00	3.88 ± 0.15
10		1.00	1.00	1.00	5.00	0.50	10.00	0.1± 0.05
11		10.00	10.00	20.00	1.00	0.50	0.00	7.01± 0.20
12		10.00	10.00	1.00	1.00	0.50	10.00	0.8 ± 0.06

Table 2. The ANOVA analysis for PB design.

Source	Coefficient	Sum of squares	df	Mean square	F-value	p-value Probe > F
Model		67.34	7	9.62	19.39	0.0062
X_1	0.82	5.47	1	5.47	11.03	0.0294
X_2	0.95	8.74	1	8.74	17.61	0.0137
X_3	1.59	26.71	1	26.71	53.83	0.0018
X_4	0.089	0.051	1	0.051	0.1	0.7648
X_5	0.8	3.64	1	3.64	7.33	0.0537
X_6	−0.33	1.1	1	1.1	2.22	0.2104
$X_2 \times X_3$	1.64	13.09	1	13.09	26.39	0.0068
Residual		1.98	4	0.5		
Cor Total		69.33	11			

$R^2 = 0.9714$; Pred $R^2 = 0.7277$; Adj $R^2 = 0.9213$; Adeq Precision = 12.497.

3.2. Optimization of media components for ethanol fermentation

PB design experiments were performed to select the critical fermentation parameters which have significant effects on ethanol production with *E. coli* FBWHR. Yeast extract, tryptone, and $Na_2HPO_4 \cdot 7H_2O$ screened out by PB design were further studied by a five-level three-factor CCD with six replications of center points to enhance the ethanol production. The design matrix of CCD, the variables and the corresponding results were shown in Table 3. The results of the second-order response surface model fitting in the ANOVA were presented in Table 4. The interactive model term yeast extract × tryptone was insignificant, because the *p*-value was greater than 0.05. All the other model terms were significant, because low *p*-values ($p < 0.05$) were obtained. Of the modeled terms, yeast extract, tryptone, $Na_2HPO_4{\cdot}7H_2O$, yeast extract × $Na_2HPO_4{\cdot}7H_2O$ and tryptone × $Na_2HPO_4{\cdot}7H_2O$ positively affected the ethanol production, while (yeast extract)^2, (tryptone)^2, and $(Na_2HPO_4{\cdot}7H_2O)^2$ negatively affected the ethanol production. The coefficient of multiple determination (R^2) was found to be 0.9898, which means that the model could explain 98.98% of the total variations in the system. The high value of the adjusted R^2 (0.9824) further supported the accuracy of the model. The predicted R^2 of 0.9517 was in the reasonable agreement with the adjusted R^2 value. Therefore, the results in terms of the production of ethanol can be illustrated by the following quadratic regression equation:

$$Y = 13.62 + 0.77\,X_1 + 1.71\,X_2 + 1.88\,X_3 + 1.23\,X_1 \times X_3 + 0.75\,X_2 \times X_3 - 2.88\,X_1^2 - 1.18\,X_2^2 - 1.53\,X_3^2 \qquad (5)$$

Where, the production of ethanol as Y is a multiple function of yeast extract, trypone and $Na_2HPO_4{\cdot}7H_2O$ concentrations.

Table 3. The CCD design for three significant variables screened by PB design in ethanol production from 30% (v/v) of hot-water sugar maple hemicellulosic wood hydrolysate by *E. coli* FBWHR.

Run	Levels	X_1, g/L	X_2, g/L	X_3, g/L	Ethanol concentration, g/L ($n = 3$)
	1.682	14.20	1.59	11.59	
	1.00	12.50	5.00	15.00	
	0.00	10.00	10.00	20.00	
	−1.00	7.50	15.00	25.00	
	−1.682	5.80	18.41	28.41	
1		7.50	5.00	15.00	5.95 ± 0.23
2		12.50	5.00	15.00	4.58 ± 0.16
3		7.50	15.00	15.00	7.7 ± 0.28
4		12.50	15.00	15.00	6.16 ± 0.11
5		7.50	5.00	25.00	5.68 ± 0.07
6		12.50	5.00	25.00	9.7 ± 0.30
7		7.50	15.00	25.00	10.88 ± 0.17
8		12.50	15.00	25.00	13.8 ± 0.12
9		5.80	10.00	20.00	3.52 ± 0.25
10		14.20	10.00	20.00	7.35 ± 0.08
11		10.00	1.59	20.00	7.04 ± 0.14
12		10.00	18.41	20.00	13.42 ± 0.12
13		10.00	10.00	11.59	6.29 ± 0.26
14		10.00	10.00	28.41	12.21 ± 0.14
15		10.00	10.00	20.00	13.25 ± 0.08

16	10.00	10.00	20.00	13.64 ± 0.23
17	10.00	10.00	20.00	14.05 ± 0.05
18	10.00	10.00	20.00	13.28 ± 0.07
19	10.00	10.00	20.00	13.52 ± 0.13
20	10.00	10.00	20.00	13.98 ± 0.10

Table 4. The ANOVA analysis and variance analysis for the quadratic response surface model using CCD design.

Parameter	Coefficient	Sum of squares	df	Mean square	F-value	*p*-value Probe > F
Intercept	13.62					
X_1	0.77	8.03	1	8.03	32.05	0.0002
X_2	1.71	39.96	1	39.96	159.53	< 0.0001
X_3	1.88	48.09	1	48.09	191.98	< 0.0001
$X_1 \times X_2$	−0.16	0.2	1	0.2	0.8	0.3907
$X_1 \times X_3$	1.23	12.13	1	12.13	48.42	< 0.0001
$X_2 \times X_3$	0.75	4.46	1	4.46	17.79	0.0018
X_1^2	−2.88	119.39	1	119.39	476.67	< 0.0001
X_2^2	−1.18	20.17	1	20.17	80.52	< 0.0001
X_3^2	−1.53	33.71	1	33.71	134.6	< 0.0001
Model		263.6	8	32.95	133.93	< 0.0001
Residual		2.71	11	0.25		
Lack of fit		2.13	6	0.35	3.07	0.1193
Pure error		0.58	5	0.12		
Cor Total		266.3	19			
CV = 5.06; R^2 = 0.9898; Pred R^2 = 0.9517; Adj R^2 = 0.9824; Adeq Precision = 30.568.						

As shown in Table 4, the second-order regression model was statistically valid given an F-value with a low probability value ($p_{model} < 0.0001$). The lack-of-fit value ($p = 0.1193$) implied the Lack of Fit was not significant relative to the pure error. The low coefficient of variation (CV = 5.06%) demonstrated that the model was precise and reliable.

The regression equations were represented graphically by three dimensional response surface plots, which were generally used to demonstrate relationships between the response and experimental levels of each variable. Therefore, the maximum production of microbial metabolites can be found by visualization of the optimum levels of each variable through the surface plots [29,30]. Figure 1 shows a response surface plot of the effect of adding yeast extract and $Na_2HPO_4{\cdot}7H_2O$ on the production of ethanol with a tryptone concentration of 10 g/L. It can be observed that the ethanol concentration increased at first and then decreased with the concentration increase of yeast extract when the concentration of $Na_2HPO_4{\cdot}7H_2O$ varied from 15 g/L to 25 g/L. The similar trend was found as the concentration of $Na_2HPO_4{\cdot}7H_2O$ increased when the concentration of yeast extract changed from 7.5 g/L to 12.5 g/L.

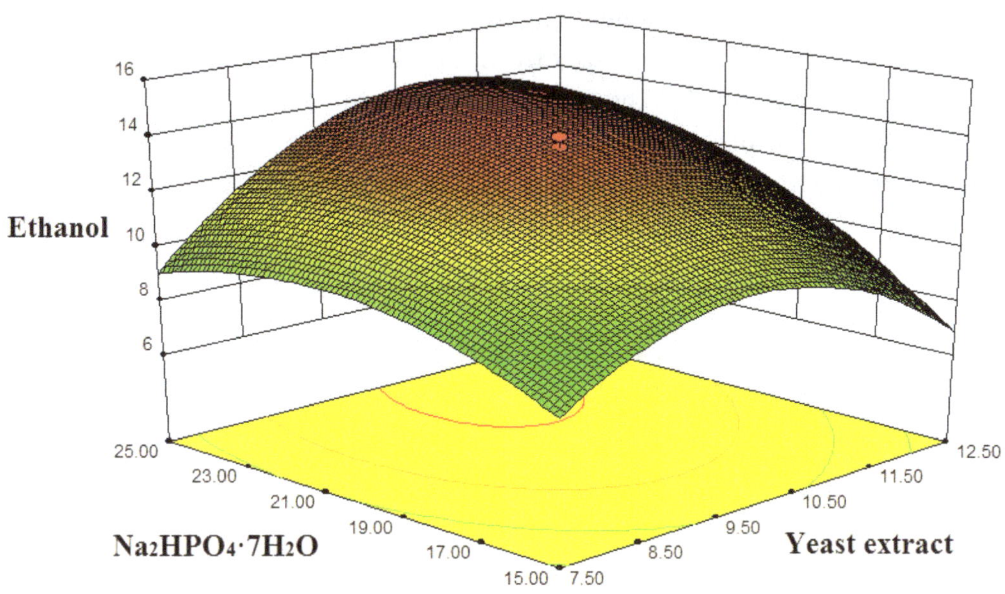

Figure 1. Response surface plot for the interaction between yeast extract and $Na_2HPO_4 \cdot 7H_2O$ on the production of ethanol with a tryptone concentration of 10 g/L.

A response surface plot that indicated the interaction between tryptone and $Na_2HPO_4 \cdot 7H_2O$ at a yeast extract concentration of 10 g/L which influenced the ethanol production was presented in Figure 2. The ethanol concentration inceased when increasing the tryptone concentration and slightly decreased at a even higher concentration level of tryptone. The ethanol production varied in a similar way as the concentration of $Na_2HPO_4 \cdot 7H_2O$ increased when the tryptone concentration varied from 5 g/L to 15 g/L.

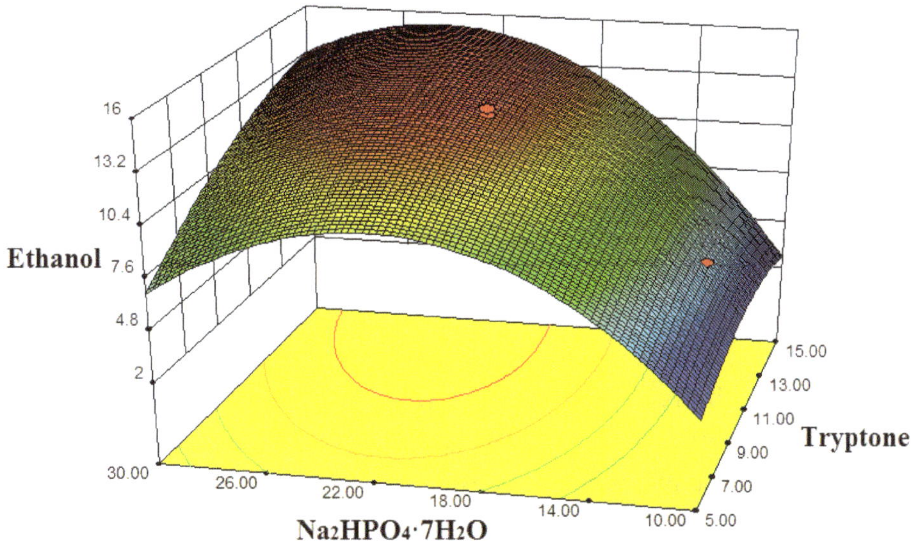

Figure 2. Response surface plot for the interaction between tryptone and $Na_2HPO_4 \cdot 7H_2O$ on the production of ethanol with a yeast extract concentration of 10 g/L.

Based on Equation 5, the optimal concentrations of yeast extract, tryptone, $Na_2HPO_4{\cdot}7H_2O$, KH_2PO_4 and NH_4Cl for achieving the maximum ethanol production from 30% (v/v) of concentrated hot-water sugar maple wood extract hydrolyzate by *E. coli* FBWHR were 10.19 g/L, 14.55 g/L, 23.21 g/L, 5 g/L and 2 g/L, respectively. The predicted maximum ethanol concentration was 15.31 g/L and the actual concentration of 15.23 ± 0.21 g/L was obtained with the optimal fermentation media, which was in close agreement to the model prediction.

3.3. Batch ethanol fermentation in 1.3 L bioreactor

Batch fermentation was conducted in a 1.3 L bioreactor for ethanol production. Utilization of individual and total sugars, biomass growth and ethanol production by recombinant *E. coli* FBWHR at 30% (v/v) of concentrated hot-water wood extract hemicellulosic hydrolyzate were depicted in Figure 3.

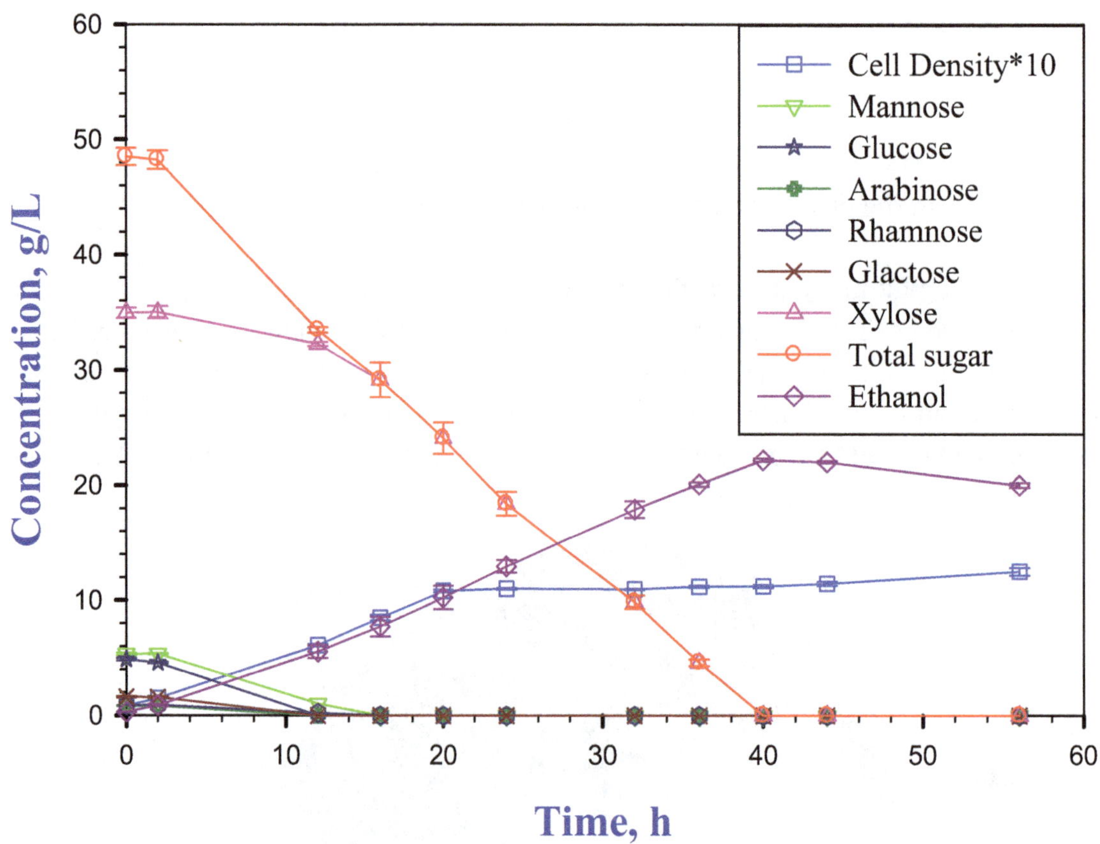

Figure 3. Utilization of individual and total sugars, biomass growth and ethanol production by recombinant *E. coli* FBWHR at 30% (v/v) of concentrated hot-water wood extract hemicellulosic hydrolyzate using optimized media in 1.3 L bioreactor.

The total sugar concentration of 48.53 ± 0.75 g/L was detected in 30% (v/v) of concentrated hot-water wood extract hydrolyzate. Under the optimal fermentation media conditions, the maximum ethanol concentration of 22.18 ± 0.13 g/L was achieved in 40 hours with a complete consumption of all the sugars. Table 5 shows the comparison of ethanol fermentation in the optimized medium to the previous study [11] (30% (v/v) of concentrated hot-water wood extract hydrolysate, recombinant *E.*

coli FBWHR). In the same fermentation process (pH, 7.0; Temperature, 35 °C, agitation, 200 rpm; air flow rate, 0.031 vvm), the maximum ethanol concentration of 22.18 ± 0.13 g/L was obtained with the optimized fermentation media, which was 1.22 times higher than that by using 20 g/L LB broth [11]. Meanwhile, the fermentation time reaching the maximum ethanol concentration was reduced from 46 hours to 40 hours. The ethanol yield and productivity increased by 1.20-fold and 1.41-fold, respectively, to the previous reported data for the same fermentation process. Owing to the sugar utilization for cell growth, cell maintenance and by-products production, the theoretical maximum ethanol yield of 0.51 g ethanol/g total sugar was not achieved.

Table 5. A comparison of ethanol fermentation by recombinant *E. coli* FBWHR using 30% (v/v) of concentrated hot-water wood extract hydrolyzate.

	This study	Wang et al. [10]
Fermentation time, hours	40	46
Ethanol concentration, g/L ($n = 3$)	22.18 ± 0.13	18.19
Ethanol yield ($Y_{p/s}$), g ethanol/g total sugar ($n = 3$)	0.46 ± 0.01	0.38
Ethanol productivity, g/(L·h) ($n = 3$)	0.56 ± 0.03	0.40

4. Conclusion

Optimization of fermentation media for ethanol production from concentrated hot-water hemicellulosic wood extract hydrolyzate by recombinant *E. coli* FBWHR has been successfully investigated combining PB design as well as CCD response surface methodology. A maximum ethanol concentration obtained in the optimized media agreed with the result of the model prediction. Batch scale-up fermentation with optimized media further enhanced the ethanol production. The fermentation time was shortened while the ethanol yield and productivity were improved. The optimal conditions studied in this work could contribute to the development of future pilot-scale ethanol production from hemicellulosic wood hydrolyzate.

Acknowledgements

NMR analysis by Mr. Dave Kiemle and ICP tests by Ms. Deb Driscoll are gratefully acknowledged. The authors are also indebted to the Bioprocess Engineering Research Group and Department of Paper and Bioprocess Engineering, SUNY ESF.

Conflict of Interest

All authors declare no conflicts of interest in this paper.

References

1. Amidon T, Wood C, Shupe A, et al. (2008) Biorefinery: Conversion of woody biomass to chemicals, energy and materials. *J Biobased Mater Bioenergy* 2: 100-120.
2. Liu S (2012) Utilization of Woody Biomass: Sustainability. *J Bioprocess Eng Biorefinery* 1: 129-139.

3. Wang Y, Liu S (2012) Butadiene production from ethanol. *J Bioprocess Eng Biorefinery* 1: 33-43.
4. Lokhorst A, Wildenborg I (2005) Introduction to CO_2 geological storage: Classification of storage options. *Oil Gas Sci Technol* 60: 513-515.
5. Balat M, Balat H (2009) Recent trends in global production and utilization of bio-ethanol fuel. *Appl Energ* 86: 2273-2282.
6. Qureshi N, Dien B, Liu S, et al. (2012) Genetically engineered *Escherichia coli* FBR5: part I. Comparison of high cell density bioreactors for enhanced ethanol production from xylose. *Biotechnol Prog* 2: 1167-1178.
7. Srirangan K, Akawi L, Moo-Young M, et al. (2012) Towards sustainable production of clean energy carriers from biomass resources. *Appl Energ* 100: 172-186.
8. Schenck A, Berglin N, Uusitalo J (2013) Ethanol from Nordic wood raw material by simplified alkaline soda cooking pre-treatment. *Appl Energ* 102: 229-240.
9. Wang Y, Liu S (2012) Pretreatment technologies for biological and chemical conversion of woody biomass. *TAPPI J* 11: 9-16.
10. Gírio F, Fonseca C, Carvalheiro F, et al. (2010) Hemicelluloses for fuel ethanol: A review. *Bioresour Technol* 101: 4775-4800.
11. Wang Y, Liu Z, Chatsko M, et al. (2013) Ethanol fermentation by *Escherichia coli* FBWHR using hot-water sugar maple wood extract hydrolyzate as substrate: A batch fermentation and kinetic study. *J Bioprocess Eng Biorefinery* 2: 1-7.
12. Tholudur A, Sorensen T, Zhu X, et al. (2005) Using Design of Experiments To Assess *Escherichia coli* Fermentation Robustness. *BioProcess Int* 3: 1-4.
13. Cockshott A, Sullivan G (2001) Improving the fermentation medium for Echinocandin B production. Part I: sequential statistical experimental design. *Process Biochem* 36: 647-660.
14. Chen P, Chiang C, Chao Y (2010) Medium optimization and production of secreted *Renilla luciferase* in Bacillus subtilis by fed-batch fermentation. *Biochem Eng J* 49: 395-400.
15. Plackett R, Burman J (1946) The design of optimum multifactorial experiments. *Biometrika* 33: 305-325.
16. Guo Y, Xu J, Zhang Y, et al. (2010) Medium optimization for ethanol production with *Clostridium autoethanogenum* with carbon monoxide as sole carbon source. *Bioresour Technol* 101: 8784-9.
17. Chauhan K, Trivedi U, Patel K (2007) Statistical screening of medium components by Plackett-Burman design for lactic acid production by *Lactobacillus sp.* KCP01 using date juice. *Bioresour Technol* 98: 98-103.
18. Mei X, Liu R, Shen F, et al. (2009) Optimization of fermentation conditions for the production of ethanol from stalk juice of sweet sorghum by immobilized yeast using response surface methodology. *Energ Fuels* 23: 487-91.
19. Pereira F, Guimarães P, Teixeira J, et al. (2010) Optimization of low-cost medium for very high gravity ethanol fermentations by *Saccharomyces cerevisiae* using statistical experimental designs. *Bioresour Technol* 101: 7856-63.
20. Grahovac J, Dodic J, Jokic A, et al. (2012) Optimization of ethanol production from thick juice: A response surface methodology approach. *Fuel* 93: 221-228.
21. Wang R, Ji Y, Melikoglu M, et al. (2007) Optimization of innovative ethanol production from wheat by response surface methodology. *Trans IChemE* 85: 404-412.

22. Yu J, Zhang X, Tan T (2009) Optimization of media conditions for the production of ethanol from sweet sorghum juice by immobilized *Saccharomyces cerevisiae*. *Biomass Bioenerg* 33: 521-526.
23. Dagnino E, Chamorro E, Romano S, et al. (2013) Optimization of the acid pretreatment of rice hulls to obtain fermentable sugars for bioethanol production. *Ind Crop Prod* 42: 363-368.
24. Cao W, Liu R (2013) Screening and optimization of trace elements supplement in sweet sorghum juice for ethanol production. *Biomass Bioenerg* 50: 45-51.
25. Liang L, Zheng Y, Shen Y (2008) Optimization of β-alanine production from β-aminopropionitrile by resting cells of *Rhodococcus sp.* G20 in a bubble column reactor using response surface methodology. *Process Biochem* 43: 758-764.
26. Thomas K, Ingledew W (1990) Fuel alcohol production: effects of free amino nitrogen on fermentation of very-high-gravity wheat mashes. *Appl Environ Microbiol* 56: 2046-2050.
27. Bakonyi P, Nemestóthy N, Lövitusz É, et al. (2011) Application of Plackett-Burman experimental design to optimize biohydrogen fermentation by *E. coli* (XL1-BLUE). *Int J Hydrogen Energ* 36: 13949-13954.
28. Suárez D, Liria C, Kilikian B (1998) Effect of yeast extract on *Escherichia coli* growth and acetic acid production. *World J Microbiol Biotechnol* 14: 331-335.
29. He J, Zhen Q, Qiu N, et al. (2009) Medium optimization for the production of a novel bioflocculant from *Halomonas sp.* V3a′ using response surface methodology. *Bioresour Technol* 100: 5922-5927.
30. Haider M, Pakshirajan K (2007) Screening and optimization of media constituents for enhancing lipolytic activity by a soil microorganism using statistically designed experiments. *Appl Biochem Biotechnol* 141: 377-390.

11

Upgrading of a wastewater treatment plant with a hybrid moving bed biofilm reactor (MBBR)

Luigi Falletti [1], Lino Conte [1], and Andrea Maestri [2]

[1] Department of Industrial Engineering, University of Padova, via Marzolo 9–35127 Padova, Italy
[2] Polesine Acque SpA, via B. Tisi da Garofolo 11–45100 Rovigo, Italy

Abstract: The wastewater treatment plant of Porto Tolle (RO, Italy) was originally projected for 2200 person equivalent (p.e.) and it was made of a pumping station, an activated sludge oxidation tank (395 m^3), a settler (315 m^3), and two sludge drying beds. Other units were not yet in use in 2008: a fine screen, a sand and grit removal unit, a new settler (570 m^3), a disinfection tank and a sludge thickener. Effective hydraulic load was 245% higher, organic load was 46% lower and nitrogen load was 39% higher than project values. Moreover, higher pollutant loads and more strict emission limits for nitrogen were expected. So the plant was upgraded: the old settler was divided into a sector of 180 m^3 that was converted into a predenitrification tank, and a sector of 100 m^3 that was converted into a hybrid MBBR tank filled with 50% AnoxKaldnes^{TM} K3 carriers; the new settler was connected to the hybrid MBBR, and the other units were started. Biofilm growth was observed two months after plant restarting, its concentration reached 1.1 g_{TS}/m^2 (0.26 kg_{TS}/m^3), while activated sludge concentration was 2.0–2.8 kg_{TSS}/m^3 in all the period of study. The upgraded plant treats 1587 m^3/d wastewater with 57 kg_{COD}/d, 23 kg_{BOD}/d and 13.3 kg_N/d, and has a significant residual capacity; the effluent respects all emission limits.

Keywords: hybrid moving bed; nitrogen; biofilm; activated sludge; wastewater treatment; upgrading

---

1. Introduction

In moving bed biofilm reactors (MBBR) the biomass grows as a biofilm on small plastic carriers that move freely into the wastewater; mixing biofilm reactors the biomass grows only on carriers, while in hybrid reactors there are both biofilm and suspended activated sludge in the same tank. Several processes with different kinds of floating carriers have been developed [1], with porous materials such as polyurethane and with nonporous material such as polyethylene. The process with the largest number of applications is the AnoxKaldnes™ MBBR with more than 400 plants in the world; it was developed in Norway, initially as a pure biofilm process, in a cooperation between the research institute SINTEF of Trondheim and the company AnoxKaldnes™ [2,3], and patented in 1991–1994. This technology is versatile and is applied for organic substance removal, nitrification and denitrification with several plant configurations; with sequencing-batch MBBR, biological phosphorus removal can be achieved [4]. Reactors for organic substance removal are usually dimensioned with surface loads as 6 kg_{BOD}/m^2d when 90% removal is requested; but for first-stage treatment of concentrated industrial wastewater, surface loads reach 30–100 kg_{COD}/m^2d [5]. Organic substance has a negative effect on nitrification, which is almost inhibited when surface load is higher than 5 g_{BOD}/m^2d [6]. Nitrification rate depends strongly on oxygen concentration (which is the limiting substrate when the ratio $O_2/N < 2$); effect of temperature decrease is negative on nitrification rate but positive effect on oxygen solubility in water [7,8]. Literature reports values of 0.8–1.0 g_N/m^2d at 15 °C and 5 mg/L O_2 with a secondary effluent [9], 0.5–0.7 g_N/m^2d at 7–15 °C and 3.5–13.3 mg/L O_2 in a plant with 8 serial MBBR, 5 of which were aerated [10], 0.5 g_N/m^2d at 6.4 °C and 6.5 mg/L O_2 in a plant with 9 serial MBBR, 4 of which were aerated [11]. With a pure oxygen tertiary MBBR, nitrification rates of 1.2–4.3 g_N/m^2d were achieved at 16–20 °C and oxygen concentrations until 20 mg/L [12]. Denitrification can be achieved with pre- or post-denitrification scheme; literature reports surface pre-denitrification rates of 0.4–0.7 g_N/m^2d at 15 °C with raw wastewater (ratio $COD/N > 3$), specific post-denitrification rates of 1.2 g_N/m^2d at 15 °C with sodium acetate as carbon source [7,10]. Some case studies of hybrid plants with AnoxKaldnes™ carriers are also reported in literature, mainly as pilot scale researches [13,14,15] and upgrading of existing activated sludge plants on basis of results of previous pilot scale experiments: Klippan in Sweden [16], Broomfield in U.S.A. [17,18], Maserà in Italy [19], Gavà-Viladecans in Spain [20]. Studies report that nitrification is significantly improved with hybrid MBBR, according to Christensson and Welander [16] ca. 85% of nitrification is due to biofilm; organic substance oxidation and sludge setteability is also improved.

This paper deals with the wastewater treatment plant of Porto Tolle (Italy), which was upgraded by dividing an existing tank into an activated sludge denitrification tank, an activated sludge oxidation tank and a hybrid MBBR oxidation tank.

2. Plant description

The wastewater treatment plant of Porto Tolle (RO, Italy) was originally projected for 2200 p.e.; expected loads were 550 m^3/d, 264 kg_{COD}/d, 132 kg_{BOD}/d and 26 kg_N/d. Emission limits were stated by the Regional Plan for Water Resanation and were: TSS ≤ 35 mg/L, COD ≤ 125 mg/L, BOD ≤ 25 mg/L, $NH_4^+ \leq 20$ mg/L, NO_2-N ≤ 1 mg/L, NO_3-N ≤ 30 mg/L, P ≤ 10 mg/L. The plant was originally made of a pumping station, a circular activated sludge oxidation tank (395 m^3) aerated by a 1480 Nm^3/h blower and fine bubble diffusers, a sedimentation unit in the external circular crown (315 m^3), and two drying beds for sludge (Figure 1A). Other units were built but not yet in use in 2008: a fine screen, a sand and grit removal unit, a circular sedimentation tank (570 m^3), a disinfection tank with peracetic acid and a sludge thickener. According to average data of 2007, the plant treated 1898 m^3/d with 35 mg/L TSS, 75 mg/L COD, 37 mg/L BOD, 19.1 mg/L TKN; hydraulic load was 245% higher, organic load was 46% lower and nitrogen load was 39% higher than project values. Wastewater was (and is) very dilute because of infiltrations of rain waters and clean water from the river Po. Even if the plant respected emission limits, the regional environmental law was near to be modified with more strict emission limits for nitrogen and the plant had no denitrification; moreover, higher pollutant incoming loads were expected for following years. So it was decided to modify the plant and to start units that were still not in use; the choice to convert part of existing volume into a hybrid MBBR was justified with the possibility of reducing activated sludge concentration in the plant. So the circular crown was divided into two sections (Figure 1B): a sector of 180 m^3 was converted into a predenitrification tank, a sector of 100 m^3 was converted into a hybrid MBBR oxidation tank. This reactor was filled with 50% AnoxKaldnes™ K3 carriers (Table 1) and aerated by a new 1000 Nm^3/h blower and medium bubbles air diffusers. The plant was modified in July 2013 and restarted with all units in August; carriers were introduced in the MBBR tank in two phases (75% in August and 25% in September).

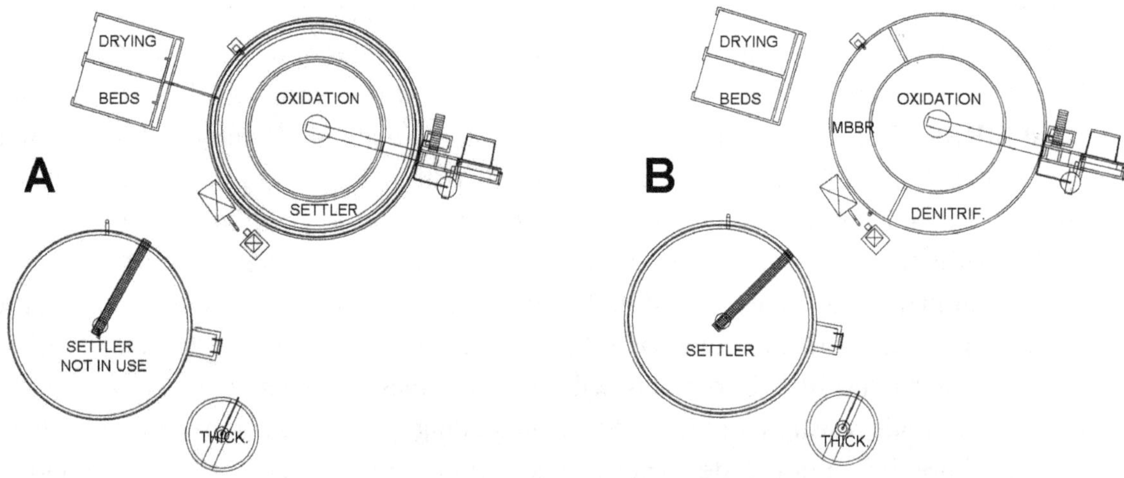

Figure 1. (A) Plant of Porto Tolle in first configuration and (B) after upgrading.

Table 1. Characteristics of AnoxKaldnes™ K3 carriers.

Shape	
Length (mm)	12
Diameter (mm)	25
Density (g/cm^3)	0.95
Nr. carriers pr. m^3	144000
Maximum filling degree	66%
Effective specific surface (m^2/m^3 carriers)	500

3. Material and methods

The upgraded plant was studied in the period 12 August–22 November 2013; hydraulic load was measured daily and samples were taken every week form raw wastewater and final effluent. These samples were analyzed to determine concentration of TSS, COD, BOD, TKN, NH_4-N, NO_2-N, NO_3-N and total nitrogen (Tot-N); analysis were conducted according to Italian standard methods published by IRSA-CNR. Since 15 September, samples of carriers were taken from the hybrid MBBR every two weeks to determine biofilm concentration, oxidizing activity and nitrifying activity with laboratory batch tests. Biofilm concentration was determined as difference of dry weight between 100 carriers with biofilm and the same carriers after cleaning with sodium hypochlorite solution. Oxidizing activity of biofilm was determined as follows: 100 carriers with biofilm were put in contact with 1 L raw wastewater under constant aeration; at the beginning of the test and after 1 hour samples were taken, filtered and analyzed to determine COD concentration; the difference between these values was divided by biofilm concentration to calculate oxidizing activity ($mg_{COD}/g_{TS}h$). Nitrifying activity was determined in a similar way, but a standard solution of ammonium chloride (ca. 50 mg/L) and sodium hydrogen carbonate (ca. 200 mg/L) was used instead of raw wastewater, and nitrifying activity ($mg_N/g_{TS}h$) was calculated from the difference in ammonium concentration divided by biofilm concentration.

4. Results and discussion

Results of analysis during restarting phase of the upgraded plant are listed in Table 2; influent pollutant concentrations were very low, mainly because of significant infiltrations of rain waters and clean water from the river Po; this dilution explains reasonably pollutant removal efficiencies which were not as high as expected. The plant respected always emission limits. Average pollutant loads were 90.5 kg_{COD}/d, 36.5 kg_{BOD}/d and 21.1 kg_N/d; since the plant was upgraded to treat up to 180 kg_{COD}/d, 90 kg_{BOD}/d and 60 kg_N/d, it has still significant residual capacity.

Table 2. Hydraulic load and pollutant concentrations at the plant of Porto Tolle.

Parameter		Raw influent			Final effluent			Removal
		min	average	max	min	average	max	
Hydraulic load	m^3/d	306	1587	2208	306	1587	2208	
TSS	mg/L	< 10	15	28	< 10	11	32	63%
COD	mg/L	< 25	57	214	< 25	25	49	56%
BOD	mg/L	< 5	23	111	< 5	6	15	74%
TKN	mg/L	6.0	13.3	41.3	< 4	< 4	< 4	85%
NH_4-N	mg/L				< 0.5	< 0.5	< 0.5	
NO_2-N	mg/L				< 0.1	< 0.1	< 0.1	
NO_3-N	mg/L				7.3	10.7	12.4	
Tot-N	mg/L	6.0	13.3	41.3	7.3	10.7	12.4	20%

Note: when values were lower than detection limits, for calculations of average values and removal efficiencies they were considered as half of detection limits.

Figure 2 shows biofilm growth in the hybrid MBBR tank, the two arrows indicate carriers introduction. There was no significant biofilm growth for almost two months after first carriers introduction (75% of total carriers amount); then biofilm concentration reached 1.1 g_{TS}/m^2 (0.27 kg_{TS}/m^3 tank) in one month; in the same period activated sludge concentration was 2.0–2.8 g_{TSS}/m^3. The start-up period of this plant was longer than the ones encountered by other two authors for similar hybrid MBBR plants fed with municipal wastewater; Di Trapani et al. [13] report that their pilot plant reached good working conditions one month after starting-up; Christennson and Welander [16] report a start-up period of 49 days for their pilot plant; Santamaria et al. [20] report a start-up period of one month. The slower biofilm growth in the hybrid MBBR of Porto Tolle can be explained with the significant dilution of raw wastewater; the plant studied by Di Trapani et al. treated an influent with 257–632 mg/L COD and 23–54 mg/L NH_4.N, and the plant studied by Christennson and Welander treated an influent with 186–256 mg/L COD and 31–37 mg/L NH_4-N. The plant studied by Santamaria et al. treated an influent with 289–481 mg/L BOD. The first biofilm growth in the full-scale hybrid MBBR plant of Maserà [19] required one month; that plant treated an influent with 210–680 mg/L COD and 40–96 mg/L TKN, and biofilm concentration reached 1.7 g_{TS}/m^2 in less than 3 months after carriers introduction.

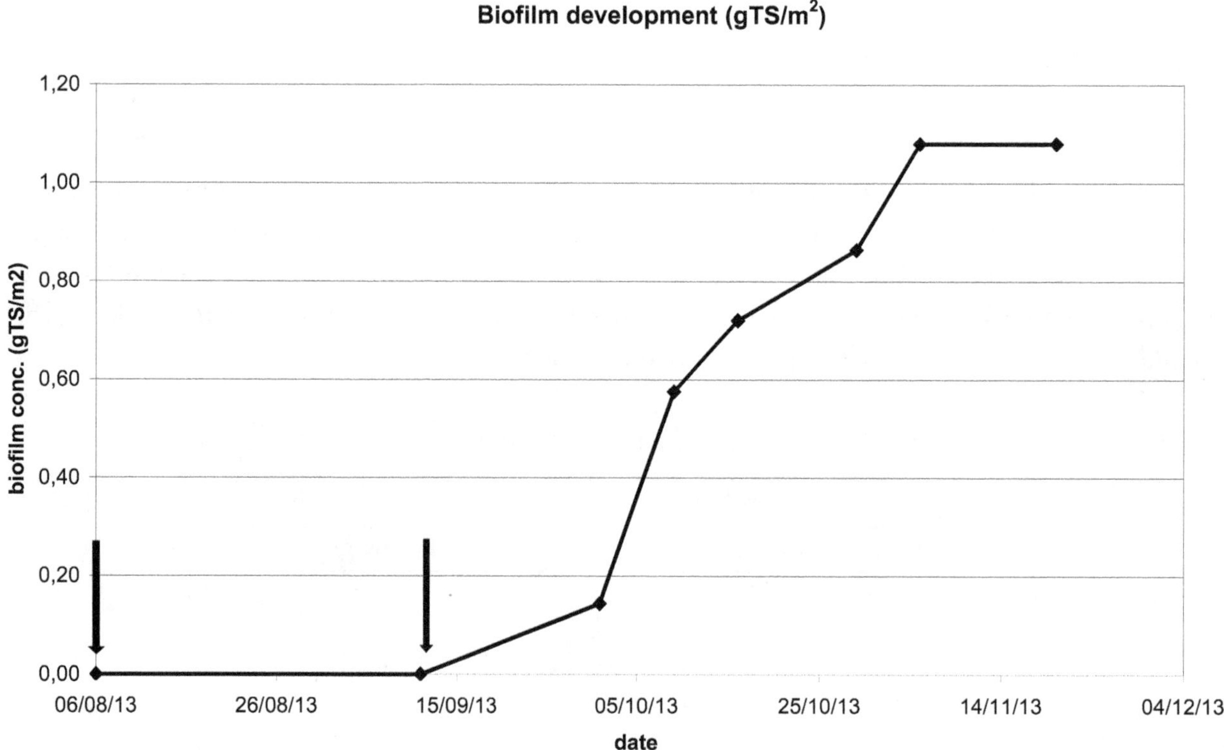

Figure 2. Biofilm development in the hybrid MBBR tank.

Figure 3 shows oxidizing activity and nitrifying activity of biofilm; significant activity (67 $mg_{COD}/g_{TS}h$ and 13.8 $mg_N/g_{TS}h$) was evidenced since the end of October. Since biofilm concentration was ca. 1.1 g_{TS}/m^2 in that period, these values can be converted into superficial removal rates, i.e. respectively 1.77 g_{COD}/m^2d and 0.36 g_N/m^2d. Christennson and Welander [16] report higher superficial nitrification rates (0.86–1.18 g_N/m^2d); also this significant difference can be explained considering the dilution of influent wastewater of Porto Tolle.

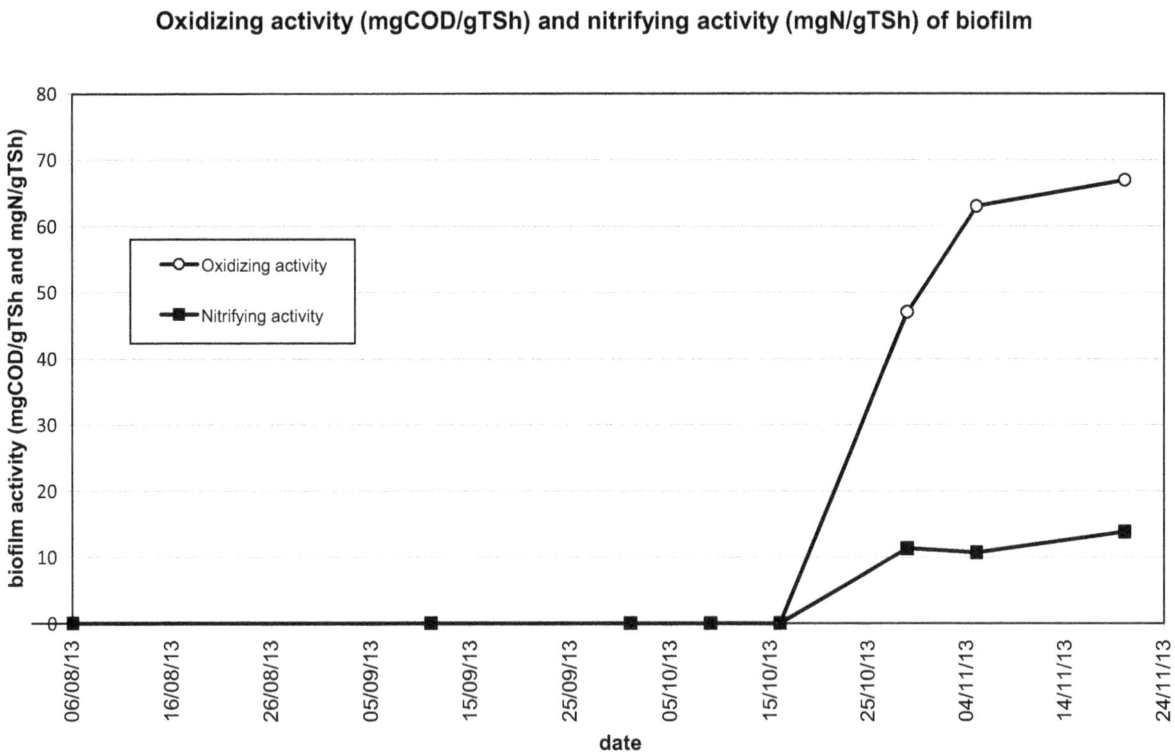

Oxidizing activity (mgCOD/gTSh) and nitrifying activity (mgN/gTSh) of biofilm

Figure 3. Oxidizing and nitrifying activity of biofilm in the hybrid MBBR tank.

5. Conclusion

These results confirm that hybrid MBBR is a suitable technology to upgrade existing activated sludge wastewater treatment plants in limited space without building new tanks. The plant of Porto Tolle was overloaded hydraulically and for nitrogen, and had no denitrification in its original scheme. So the old settler was converted into a predenitrification tank and a hybrid MBBR tank, and another settler (already built but not yet in use in 2008) was connected to the biological section. The plant was so upgraded to treat up to 180 kg_{COD}/d, 90 kg_{BOD}/d and 60 kg_N/d; the same biological section without carriers could treat up to 120 kg_{COD}/d, 60 kg_{BOD} and 40 kg_N/d with current emission limits.

During the start-up phase the plant received very diluted wastewater with 57 mg/L COD, 23 mg/L BOD and 13.3 mg/L TKN on average basis. Thus pollutant removal efficiencies were apparently low even if the plant respected emission limits. First biofilm growth was observed ca. two months after first carriers introduction, and its concentration reached 1.1 g_{TS}/m^2 in the following month. Start-up time was longer and biofilm concentration was lower than respective values reported by other authors for similar hybrid MBBR plants, reasonably because of very low BOD and TKN concentration in incoming wastewater. Laboratory batch tests on biofilm evidenced an oxidizing activity of 67 $mg_{COD}/g_{TS}h$ (1.77 g_{COD}/m^2d) and a nitrifying activity of 13.8 $mg_N/g_{TS}h$ (0.36 g_N/m^2d).

The plant has a significant residual capacity (both for organic substance oxidation and for total nitrogen removal) that will be useful for future pollutant load increases or if emission limits become stricter than current ones.

Conflict of Interest

All authors declare no conflicts of interest in this paper.

References

1. Pastorelli G, Processi a biomassa adesa a letto mobile. Sviluppi nelle tecniche di depurazione delle acque reflue (Moving bed biofilm processes, in "Development in wastewater treatment techniques"); 2000. 52nd Sanitary-Environmental Engineering Course proceedings, Milan.
2. Ødegaard H, Rusten B, Siljudalen JG (1999) The development of the moving bed biofilm process—from idea to commercial product. *Eur Water Manage* 2: 6-43.
3. Helness H, Ødegaard H (2001) Biological phosphorus and nitrogen removal in a sequenching batch moving bed biofilm reactor. *Water Sci Technol* 43: 233-240.
4. Ødegaard H, Rusten B, Westrum T (1994) A new moving bed biofilm reactor—Application and results. *Water Sci Technol* 28: 157-165.
5. Ødegaard H, Gisvold B, Strickland J (2000) The influence of carrier size and shape in the moving bed biofilm process. *Water Sci Technol* 41: 383-391.
6. Rusten B, Hem L, Ødegaard H (1995) Nitrification of municipal wastewater in moving-bed biofilm reactors. *Water Environ Res* 67: 75-86
7. Ødegaard H, Rusten B (1993) Norwegian experiences with nitrogen removal in a moving bed biofilm reactor. *Documentation of* 9: 205-221.
8. Salvetti R, Azzelino A, Canziani R, et al. Effects of temperature on tertiary nitrification in moving-bed biofilm reactors. *Water Res* 40: 2981-2993.
9. Hem LJ, Rusten B, Ødegaard H (1994) Nitrification in a moving bed biofilm reactor. *Water Res* 28: 1425-1433.
10. Rusten B, Hem LJ, Ødegaard H (1995) Nitrogen removal from dilute wastewater in cold climate using moving-bed biofilm reactors. *Water Environ Res* 67: 65-74.
11. Rusten B, Siljudalen JG, Bungun S (1995) Moving bed biofilm reactors for nitrogen removal: from initial pilot testing to start-up of the Lillehammer WWTP. *WEFTEC Proceedings* 95: 21-25.
12. Bonomo L, Pastrorelli G, Quinto E, et al. (2000) Tertiary nitrification in pure oxygen moving bed biofilm reactors. *Water Sci Technol* 41: 361-378.
13. Di Trapani D, Mannina G, Torregrossa M, et al. (2008) Hybrid moving bed biofilm reactors: a pilot plant experiment. *Water Sci Technol* 57: 1539-1545.
14. Di Trapani D, Mannina G, Torregrossa M, et al. (2010) Comparison between hybrid moving bed biofilm reactor and activated sludge system: a pilot plant experiment. *Water Sci Technol* 61: 891-902.
15. Di Trapani D, Christensson M, Ødegaard H (2011) Hybrid activated sludge/biofilm process for the treatment of municipal wastewater in a cold climate region: a case study. *Water Sci Technol* 63: 1121-1129.
16. Christensson M, Welander T (2004) Treatment of municipal wastewater in a hybrid process using a new suspended carrier with large surface area. *Water Sci Technol* 49: 207-214.
17. Josslin B, Johnson CH, Haegh M. Increasing BNR capacity in activated sludge systems by use of the HYBAS™ process—Example from a full scale installation at Broomfield BWRF in Colorado; 2006. WEFTEC Annual Conference & Exposition, Dallas, USA.

18. Rutt K, Seda J, Johnson CH (2006) Two year case study of integrated fixed film activated sludge (IFAS) at Broomfield, CO WWTP. *Proceedings of the water environment federation* 13: 225-239.
19. Falletti L, Conte L (2007) Upgrading of activated sludge wastewater treatment plants with hybrid moving-bed biofilm reactors. *Ind Eng Chem Res* 46: 6656-6660.
20. Santamaria A, Zalakain G, Hernández M, et al. (2011) Remodelación del tratamiento biológico de la EDAR Gavà-Viladecans mediante el proceso híbrido Hybas (Modification of biological treatment of EDAR Gavà-Viladecans with hybrid process Hybas, in Spanish language). *Agua* 327: 58-65.

12

DDTs, PCBs and PBDEs contamination in Africa, Latin America and South-southeast Asia—a review

Peter Mochungong and Jiping Zhu

Exposure and Biomonitoring Division, Health Canada, Ottawa, Canada K1A 0K9

Abstract: Levels of polybrominated biphenyl ethers (PBDEs), polychlorinated biphenyls (PCBs), and dichlorodiphenyltrichloroethane and its degradation products (DDTs) in the environment (ambient air, soil and aquatic mammals) and in humans from the developing regions (Africa, Latin America, and South-southeast Asia) are reviewed. Higher DDTs levels in certain parts of the developing regions due to agricultural applications and disease control measures are evident. The data however do not indicate higher levels of PCBs and PBDEs in the developing regions compared to developed countries. We also compared globally the levels of these chemicals in human milk sampled since year 2000. Human milk data again showed higher DDTs levels in the developing regions. For PBDEs, though current levels in human milk from the developing regions do not exceed levels found in the developed countries, data suggest the levels of PBDEs in the developing regions may be on the rise.

Keywords: persistent organic pollutants; developing regions; developed regions; trends; breast milk

1. Introduction

Polychlorinated biphenyls (PCBs), DDTs, which include dichlorodiphenyltrichloroethane (DDT) and its degradation products dichlorodiphenyldichloroethylene (DDE) and dichlorodiphenyldichloroethane (DDD), and polybrominated diphenyl ethers (PBDEs) are persistent organic pollutants (POPs) and are bioaccumulative. They can be transported over long distances. These chemicals also show toxicological responses in human and wildlife. PCBs and DDTs are commonly referred to as legacy POPs, while PBDEs as emerging POPs [1].

PCBs act in humans on multiple organs and organ systems including the liver, kidneys, pancreas, immune and nervous system functions and sex steroid hormonal systems, leading to increased risk of cancer, diabetes, liver disease, infertility, ischemic heart disease and hypertension [2]. A weak epidemiologic association has been identified between low-level perinatal exposure to PCBs and effects on cognitive and motor development of children [3]. Co-planar PCBs exhibit teratogenic properties and are also endocrine disrupters [4]. Gestational and lactational exposures to co-planar PCBs have been associated with impaired cognitive development and behavioural problems [5]. Commercial use of PCBs started around 1929 and steadily increased until 1979 when production and sales were banned in the United States. The substances were subsequently banned in many other developed countries.

DDTs belong to a group of organochlorine pesticides (OCPs) that have been used extensively in agriculture and for controlling disease vectors. Experimental and epidemiologic evidence have associated DDTs with various cancers such as those of the pancreas and liver as well as with disruption of the endocrine system [6]. In a United States study, DDTs have been associated with premature births [7]. Chronic, low level exposures to DDTs have been associated with a broad range of non-specific symptoms such as headache, dizziness, fatigue, weakness, nausea, chest tightness, difficulty breathing, insomnia, confusion, and concentration difficulties [8]. Occupational exposure to DDTs has been associated with reduced verbal attention, visuomotor speed, and with increased neuropsychological and psychiatric symptoms among retired workers aged 55–70 years in Costa Rica [9]. DDTs were banned in developed countries in the 1970s and 1980s due to their negative impact on non-target organisms and bioaccumulative potential in biota and humans [10,11].

PBDEs are brominated flame-retardants that have been widely used in upholstered furniture containing polyurethane foam, in insulation material for wires and cables and in high impact polystyrene in electronics and computers [12]. PBDEs have endocrine disrupting effects [13,14]. Because of their ubiquitous presence, bioaccumulation and potential toxicities to wildlife and humans, PBDEs (excluding the DecaBDE technical mixture) have been added to the list of POPs in the Stockholm Convention [15]. The European Union imposed a ban in 2004 on the production, use and import of PentaBDE and OctaBDE. In the United States, efforts are underway to phase out DecaBDEs by all current producers by the end of 2013.

Regulations now exist for many legacy and emerging POPs, and monitoring spatial and temporal changes is important to follow the results of regulation as well as for monitoring exposure risks and linking these to possible effects [16]. Compared to countries in the developed region, monitoring data of POPs remains inadequate in the developing regions. While informative reviews on countries and regions in the developed region have been conducted [15-18], a systematic review of POPs in developing regions as a whole is lacking. The main objective of this paper is to carry out a comprehensive review on the levels of aforementioned POPs in ambient air, soil and aquatic mammals as well as humans from developing regions, and compare those with countries in the developed region to provide a global perspective on the contamination of these POPs. The developing regions included are Africa, Latin America, and South-Southeast (S-SE) Asia. POPs in China was not included in this review as there are already several published reviews on the subject from China, including a recent review on levels of POPs in general population in China and the development of management policies in China to deal with the pollutants [19]. POPs contamination in China's water has also been reviewed [20].

Figures constructed for this manuscript are presented according to how they are discussed. In some cases, more than one study from a single country is used to demonstrate either spatial or temporal, or both, changes. In order to compare levels reported from different studies, only the central values that are reconstructed from each study are presented in all figures. A central value (dotted in diamond form) is an arithmetic mean (AM), a geometric mean (GM) or a median value. Some studies only reported range of concentrations, in this case 1/3 of the range was used as the central value, as most of the environmental data is log-normal distributed and 1/3 is roughly in the middle of a log scale. Levels in soil are ng/g dry weight (dw) and in ambient air is ng/m^3. For aquatic mammals and human blood and milk, values (ng/g lipid weight (lw)) are on the basis of sample content of extractable lipids. Levels of POPs in all figures are shown in log scale.

Levels of DDTs, PCBs and PBDEs are summarized in figures 1, 2 and 3 respectively. Each figure is further divided into a, b and c for different environmental matrices. Levels of POPs in human milk are summarized in Figure 4. Different colors are used in Figure 4: blue for developed regions, red for Africa, green for S-SE Asia, and purple for Latin America. The references for all the data used in the figures are mentioned in the text. We selected soil, ambient air and aquatic mammal as three representative media to evaluate POPs in the environment. Data on ambient air and soil from the developing regions are relatively available compared to other environmental media; besides, they are also important exposure routes to the general population. Aquatic mammals are apex predators, with long life expectancy rates; thus making them sentinel or indicator species for environmental perturbations [21]. No specific time period was applied during the data collection phase. Nonetheless, breast milk data focused on post 2000 samples in an effort to give current situation of human exposure. This review focused mainly on spatial comparisons, though temporal comparisons are also considered when data are available. Among OCPs, only DDTs is covered because it is increasingly relied upon in many regions of Africa and Asia where malaria remains a problem [22].

2. Levels of POPs in the environment in developing regions

2.1. DDTs

Levels of DDTs in soil from Senegal and Gambia [23], Congo [24], Tanzania [25], Uganda [26,27], and Ethiopia [28] in Africa; Brazil [29] and Chile [30,31] in Latin America; and Vietnam [32] and India [33] in S-SE Asia are summarized in Figure 1a. In general, levels of DDTs in soil samples from both high and low DDT application areas were in the range of 10–1,000 ng/g, with exceptions of background soil samples from Uganda (< 1 ng/g) and with some soil samples from Chile (> 1,000 ng/g). The higher levels in Chile and India can be attributed to the fresh application of DDT to curb the spread of malaria [30,31]. For comparison, levels from Canada [34] and from the Corn Belt of the United States [35] were at 10 ng/g and less (Figure 1a).

Low application and high application areas reflect the intensity of use of DDT and this intensity can vary significantly between countries, between different areas within a country, as well as between different regions of the world. Examples of high application areas include those where DDT was used against pests in agriculture, for the control of malaria (for example through indoor residual spraying programs), and former pesticide storage areas because of possible leaching of DDTs to the soil. Low application areas are the general urban, rural and background areas. Two studies in Gambia [23] and Tanzania [25] provided clear evidence of associations between the levels of DDTs in the soil and the intensity of DDT use by observing the following concentration gradient: the highest levels were found in DDT storage areas, followed by agriculture areas with large scale of DDT application and then by city farms with some degree of DDT use, the lowest levels were reported in soil samples from background areas where there was no DDT use.

While some POPs such as PCBs and PBDEs showed strong urban-rural gradients in soil contamination [36], levels of DDTs did not show such an urban-rural gradient due to the presence of local sources of the chemical in rural areas [34]. However, in general DDTs in soil of urban areas were 1–2 orders of magnitude higher than those in rural areas when rural sources were excluded [37,38].

DDTs in ambient air are presented in Figure 1b. Data are compiled from Congo [39], Malawi [40] and Zimbabwe [41] in Africa; southern Mexico, the Brazilian mountains and Costa Rica [41,42,43] in Latin America; and India, Vietnam, Thailand and the Solomon Islands in S-SE Asia [44]. Central values of DDTs in ambient air samples from urban areas were in the range of 1–10 ng/m^3, while less than 0.1 ng/m^3 was found in rural areas. This correlates with the difference in DDTs levels in rural and urban soils as we discussed earlier. Soils act as a continuing emission source of atmospheric contamination through volatilization [39]. Air samples from the remote Brazilian mountains had the lowest levels of DDTs (< 0.1 ng/m^3). Presence of this low background level could be largely due to long range transport of DDTs from other areas, although some unknown local sources might be present too.

Levels of DDTs in ambient air samples from several countries in the developed region such as Japan [45], Canada [46,47], Australia, Sweden and the Baltic Island [41], were also included in the figure and levels in these developed countries were generally lower at 0.01–0.1 ng/m^3. Most of the developing countries described in this review are located in the tropical regions where climatic conditions favor dissipation of DDT and other chlorinated pesticides. Continuing use of DDT in indoor residual spraying programs in some parts of the region and emissions from old pesticide storage areas may be attributed to the higher DDTs levels in the developing regions [37].

Central values of DDTs in aquatic mammals from Argentina [48], Brazil [49], East Greenland [50], Hawaiian Islands [51], India [52,53], McMurdo Sound-Antarctica [54], Norway [50], Russia [55,56,57], South Africa [58], Tanzania [59], Western Sahara [60] are summarized in Figure 1c. Levels of DDTs in the developing regions (100–100,000 ng/g) varied quite a lot and the upper levels (100,000 ng/g) were an order of magnitude higher than those in the developed countries (10,000 ng/g). For aquatic mammals, many factors including species, age, and gender and tissue type can influence the level of DDTs [61]. While DDE was reported to be the major congener of DDTs in the developed regions, most of the studies from the developing regions reported higher levels of the parent compound (DDT) compared to its metabolites (DDE or DDD) in biota. High levels of DDTs in Brazil and India could be attributed to the fact that the sampled aquatic mammals inhabited riverine and estuarine ecosystems that were in close proximity to pollution sources [49,52]. Levels of DDTs in India however, showed a 3- to 10- fold decrease between 1993–1996 and 2000–2001. This corresponds to the decrease in use of DDT from 19,000 tons/year in the early 1990s to 7,000 tons in 2001 to 2002 [49]. Although levels of DDTs in samples from South Africa and Western Sahara fringe were similar to those from developed countries, the former samples contained higher DDT to DDE ratios suggesting fresh DDT applications in the area [60].

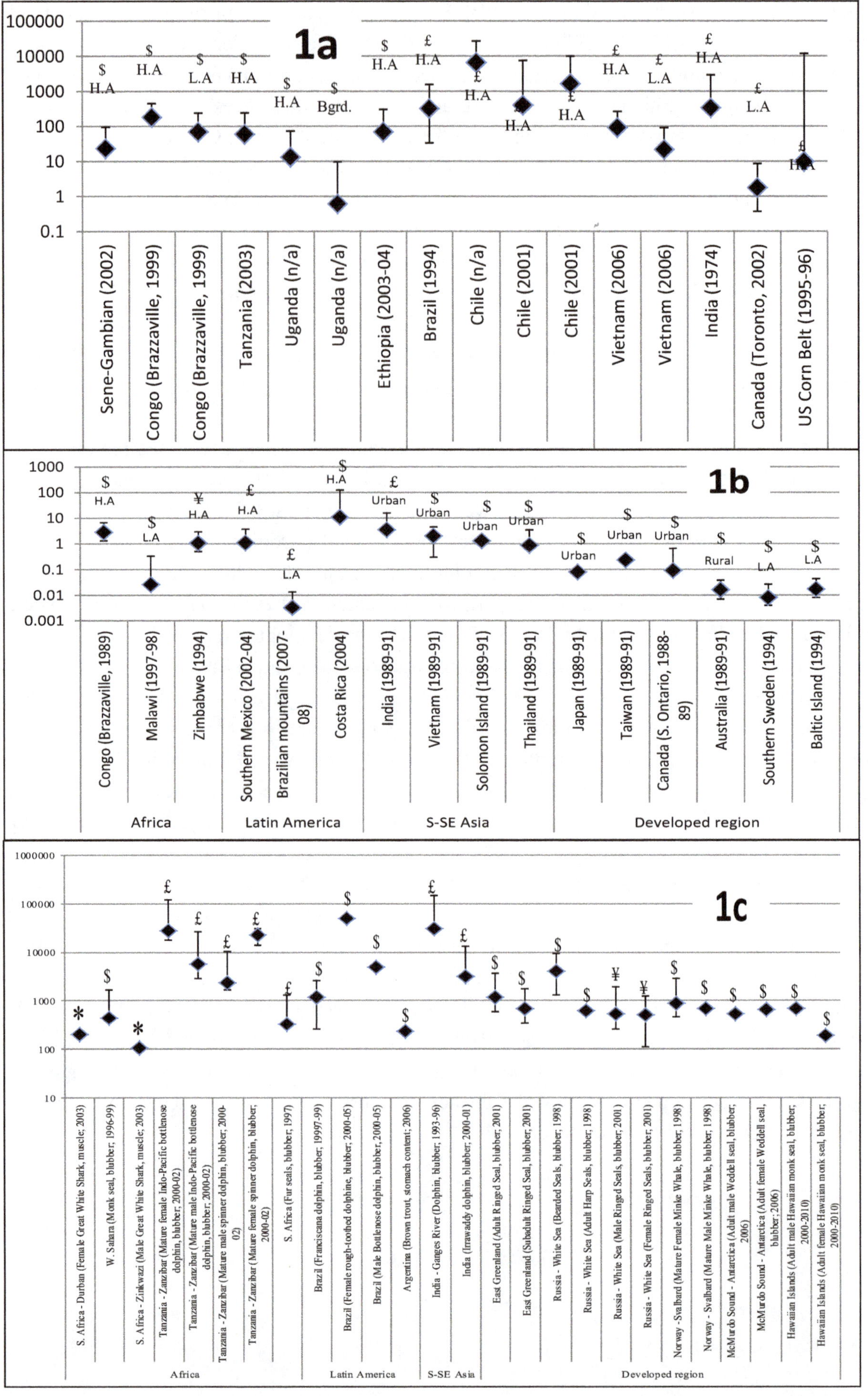
1a
100000
10000
1000
100
10
1
0.1
Sene-Gambian (2002)
Congo (Brazzaville, 1999)
Congo (Brazzaville, 1999)
Tanzania (2003)
Uganda (n/a)
Uganda (n/a)
Ethiopia (2003-04)
Brazil (1994)
Chile (n/a)
Chile (2001)
Chile (2001)
Vietnam (2006)
Vietnam (2006)
India (1974)
Canada (Toronto, 2002)
US Corn Belt (1995-96)
1b
1000
100
10
1
0.1
0.01
0.001
Congo (Brazzaville, 1989)
Malawi (1997-98)
Zimbabwe (1994)
Southern Mexico (2002-04)
Brazilian mountains (2007-08)
Costa Rica (2004)
India (1989-91)
Vietnam (1989-91)
Solomon Island (1989-91)
Thailand (1989-91)
Japan (1989-91)
Taiwan (1989-91)
Canada (S. Ontario, 1988-89)
Australia (1989-91)
Southern Sweden (1994)
Baltic Island (1994)
Africa
Latin America
S-SE Asia
Developed region
1c
1000000
100000
10000
1000
100
10
S. Africa - Durban (Female Great White Shark, muscle; 2003)
W. Sahara (Monk seal, blubber; 1996-99)
S. Africa - Zinkwazi (Male Great White Shark, muscle; 2003)
Tanzania - Zanzibar (Mature female Indo-Pacific bottlenose dolphin, blubber; 2000-02)
Tanzania - Zanzibar (Mature male Indo-Pacific bottlenose dolphin, blubber; 2000-02)
Tanzania - Zanzibar (Mature male spinner dolphin, blubber; 2000-02)
Tanzania - Zanzibar (Mature female spinner dolphin, blubber; 2000-02)
S. Africa (Fur seals, blubber; 1997)
Brazil (Franciscana dolphin, blubber; 19997-99)
Brazil (Female rough-toothed dolphine, blubber; 2000-05)
Brazil (Male Bottlenose dolphin, blubber, 2000-05)
Argentina (Brown trout, stomach content; 2006)
India - Ganges River (Dolphin, blubber; 1993-96)
India (Irrawaddy dolphin, blubber; 2000-01)
East Greenland (Adult Ringed Seal, blubber; 2001)
East Greenland (Subadult Ringed Seal, blubber; 2001)
Russia - White Sea (Bearded Seals, blubber; 1998)
Russia - White Sea (Adult Harp Seals, blubber; 1998)
Russia - White Sea (Male Ringed Seals, blubber; 2001)
Russia - White Sea (Female Ringed Seals, blubber; 2001)
Norway - Svalbard (Mature Female Minke Whale, blubber; 1998)
Norway - Svalbard (Mature Male Minke Whale, blubber; 1998)
McMurdo Sound - Antarctica (Adult male Weddell seal, blubber; 2006)
McMurdo Sound - Antarctica (Adult female Weddell seal, blubber; 2006)
Hawaiian Islands (Adult male Hawaiian monk seal, blubber; 2000-2010)
Hawaiian Islands (Adult female Hawaiian monk seal, blubber; 2000-2010)
Africa
Latin America
S-SE Asia
Developed region

Figure 1. Levels of DDTs from countries in the developed and developing regions: (a) levels in soil, ng/g dry weight; (b) levels in ambient air, ng/m^3; (c) levels in aquatic mammals, ng/g lipid weight. Year of sample collection is indicated in bracket. Central values ($ = AM; £ = 1/3 range; ¥ = GM; € = Median; * = single data point) are presented. The bar indicates minimum and maximum range if reported in the literature. H.A = High Application; L.A = Low Application; Bgrd = Background. S-SE Asia = South and Southeast Asia.

2.2. PCBs

Many countries in the developing regions have several PCB sources such as uncontrolled (open) burning of municipal waste, PCB-containing waste imported from developed countries and the use of PCBs in condensers and transformers [62]. So far, environmental monitoring of PCBs in the developing regions is not well developed; only limited data are available. Central values of PCBs in soil collected in urban and industrial areas from South Africa [62,63], Chile [64] and Brazil [65] were in the range of 1–10 ng/g, while the levels in Vietnam [66,67] were between 10 and 100 ng/g (Figure 2a). A general concentration gradient of PCBs in soil (industrial/urban area > agricultural area > background area) was observed in the cases of South Africa [62,63] and Vietnam [66]. This can be expected as samples were collected from urban and industrial areas that are characterized by diverse sources of PCBs previously alluded to. Shallow soils (1–2 cm in depth) were also found to contain lower levels of PCBs than the surface soils (0–0.5 cm in depth), due to possible higher levels of organic carbon in surface soils and atmospheric deposition [62]. On the contrary, lower PCB levels (0.1–1 ng/g) were observed in background areas from Africa, South America and Asia [68].

Background levels of PCBs from Europe and North America however, were about 10 times higher than, while levels in Australia were similar to, the background levels in the developing regions [68]. This is likely due to the fact that countries in the northern hemisphere consumed an estimated 97% of globally produced PCBs between 1930 and1993 [69].

Figure 2b illustrates levels of PCBs in ambient air. Data were collected from Ghana and South Africa [62], the Ivory Coast, Gambia and Cape Verde [70], Southern Mexico [42] and Chile [71], Vietnam, India and the Solomon Islands [41]. Data from countries in the developed region such as Switzerland [72], United Kingdom [73], Japan [74], Spain [75] and the Great Lakes area and Chicago in the United States [76] were also included in the figure. Ambient levels of PCBs in the developing regions were either lower than (in the case of Africa and Latin America) or similar to (in the case of S-SE Asia) those found in countries in the developed region.

The higher PCB levels were found in ambient air from S-SE Asian region (Vietnam, India and The Solomon Islands) than other two developing regions. This echoes the relatively high PCB levels found in soil from countries in S-SE Asian region like Vietnam (Figure 2a). It may also be due to the fact that the air samples from Vietnam, India and Solomon Islands were collected a decade earlier (1989–1991) than the other air samples. The extensive use of PCB transformers and capacitors in the S-SE Asian region might be a contributing factor. Dielectric oils containing PCBs are purportedly still widely used in some countries within the developing region [66,67]. In general, PCB levels in urban areas were higher than those in the rural areas. This can be attributed to the production and application of PCBs in urban areas [77,78]. One exception is however the data from Southern Mexico (2002–2004) and India (1994) (Figure 2b), where PCB levels were higher in rural areas. This reversed situation may be due to episodic emissions from non-point sources such as burning of municipal waste, which are common practice in rural areas in the tropical countries [79].

Levels of PCBs in the same or closely related species of aquatic mammals from Argentina [48], Brazil [80,81], Canada [50,82], India [52,53,83], Mauritius[61], Russia [56,84], USA-Florida Coast [85], Western Sahara [60] are shown in Figure 2c. All except one study reported tissue levels in blubber. Blubber is the primary storage location of fat in most aquatic mammals and POPs such as PCBs and DDTs show positive correlation with fat content of the mammals [48]. Generally, PCB levels in aquatic mammals in Africa were in the range of 100–1,000 ng/g, in Latin America were 10–10,000 ng/g, and in S-SE Asia were 100–10,000 ng/g. In comparison, levels were one to two orders of magnitude higher (1,000–100,000 ng/g) in countries in the developed region. The 10 times higher levels of PCBs in aquatic mammals from Florida coast in the United States than those in the neighboring Latin America exemplify higher PCBs levels in aquatic mammals from the developed regions. Very high levels in aquatic mammal samples from Vietnam compared to the other countries could be attributed to the usage of electrical equipment containing PCBs and the use of PCBs in artillery and other chemical weapons during the second Indo-Chinese war (1961–1971) [86].

Despite the fact that PCBs have been banned in developed countries for more than 30 years, levels of PCBs in soil, air and aquatic mammals from developed countries were generally higher than those in the developing region. Especially, aquatic mammals at the top of the food chain are yet to show meaningful declining levels [82]. This might be as a result of heavy use of PCBs in the developed regions in the 1960s and 1970s.

2.3. PBDEs

Data on PBDEs in environmental media from Africa, S-SE Asia, and Latin America are very limited, particularly for their levels in soil. As a result, we were not able to construct a summary figure for levels of PBDEs in soil. Only the levels in ambient air and in aquatic vertebrates are presented and discussed in this section.

PBDEs in ambient air from Argentina [87], Botswana [87,88], Costa Rica [88], India [89], South East Asian Sea [90] and Vietnam [90] are presented in Figure 3a. For comparison, levels of PBDEs in England and Ireland [91], the United States [78], East Greenland Sea [90], South West Greenland [93], Norway–Svalbard [87] and Canadian and Russian Arctic [17] are included in the figure as well. With the exception of the higher levels in the Canadian Arctic, which was collected in the early 1990s, PBDEs levels in ambient air were in the range of 0.001–0.1 ng/m^3 and there was no clear spatial difference between developing regions and countries in the developed region. Offshore atmospheric levels (Southeast Asian Sea, East Greenland Sea and South West Greenland) were in the range of 0.0001–0.001 ng/m^3, which were one to two orders of magnitude lower than the levels on land [90].

It is important to note that data from Botswana study showed a possible increase in atmospheric PBDE levels from year 2004 (0.006 ng/m^3) to 2007 (0.04 ng/m^3) in the region [87,88]. Some Latin American and S-SE Asian countries are primary destinations for electronic wastes (e-wastes) from Europe and North America for recycling, which may attribute to the levels of atmospheric PBDEs in these regions [89]. Local burning of municipal waste can also be a major contributor to atmospheric PBDEs, particularly in the non- source regions like the Arctic and Subarctic communities in Canada [17].

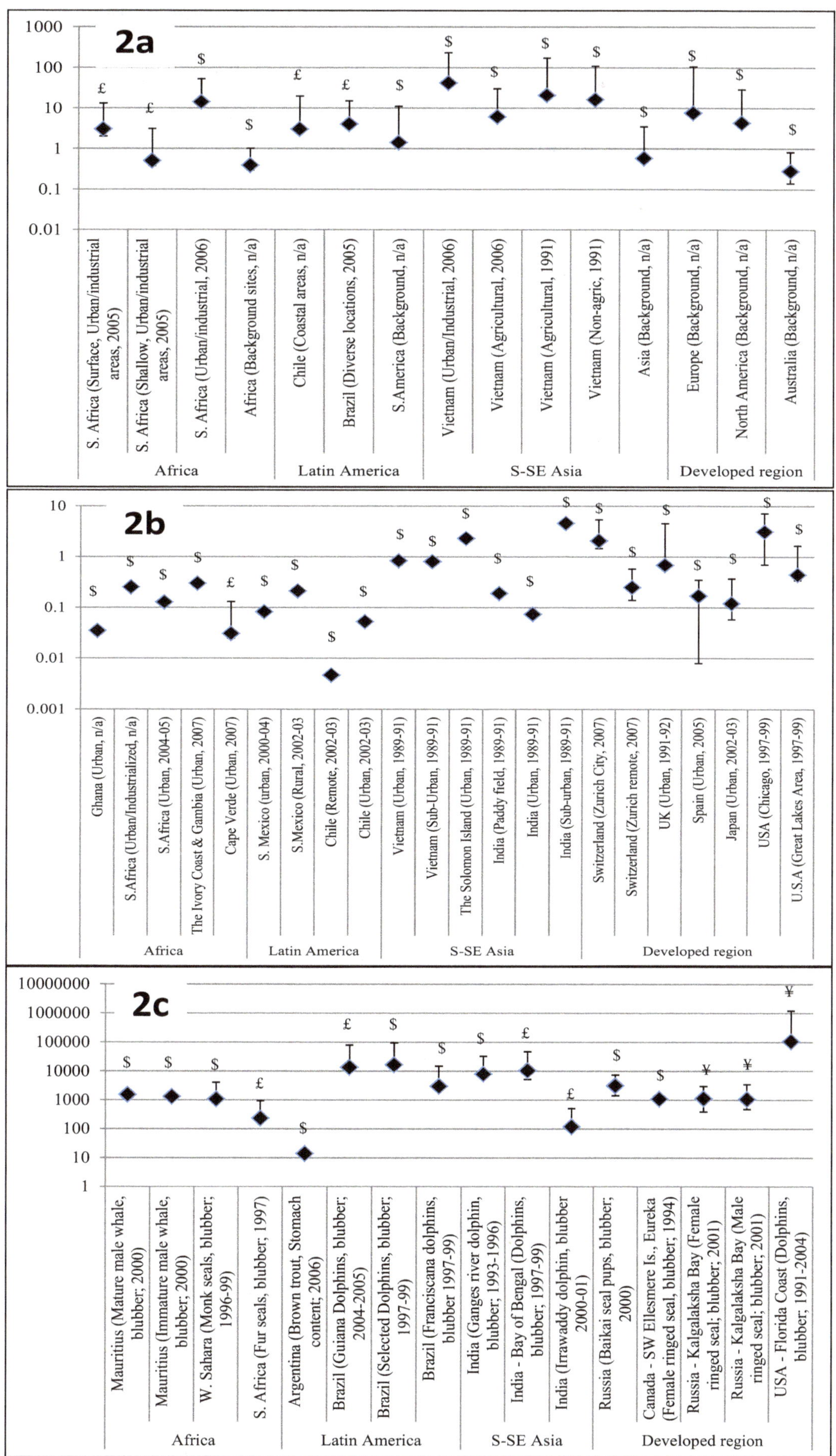
2a
1000
100
10
1
0.1
0.01
S. Africa (Surface, Urban/industrial areas, 2005)
S. Africa (Shallow, Urban/industrial areas, 2005)
S. Africa (Urban/industrial, 2006)
Africa (Background sites, n/a)
Chile (Coastal areas, n/a)
Brazil (Diverse locations, 2005)
S.America (Background, n/a)
Vietnam (Urban/Industrial, 2006)
Vietnam (Agricultural, 2006)
Vietnam (Agricultural, 1991)
Vietnam (Non-agric, 1991)
Asia (Background, n/a)
Europe (Background, n/a)
North America (Background, n/a)
Australia (Background, n/a)
Africa
Latin America
S-SE Asia
Developed region
2b
10
1
0.1
0.01
0.001
Ghana (Urban, n/a)
S.Africa (Urban/Industrialized, n/a)
S.Africa (Urban, 2004-05)
The Ivory Coast & Gambia (Urban, 2007)
Cape Verde (Urban, 2007)
S. Mexico (urban, 2000-04)
S.Mexico (Rural, 2002-03
Chile (Remote, 2002-03)
Chile (Urban, 2002-03)
Vietnam (Urban, 1989-91)
Vietnam (Sub-Urban, 1989-91)
The Solomon Island (Urban, 1989-91)
India (Paddy field, 1989-91)
India (Urban, 1989-91)
India (Sub-urban, 1989-91)
Switzerland (Zurich City, 2007)
Switzerland (Zurich remote, 2007)
UK (Urban, 1991-92)
Spain (Urban, 2005)
Japan (Urban, 2002-03)
USA (Chicago, 1997-99)
U.S.A (Great Lakes Area, 1997-99)
Africa
Latin America
S-SE Asia
Developed region
2c
10000000
1000000
100000
10000
1000
100
10
1
Mauritius (Mature male whale, blubber; 2000)
Mauritius (Immature male whale, blubber; 2000)
W. Sahara (Monk seals, blubber; 1996-99)
S. Africa (Fur seals, blubber; 1997)
Argentina (Brown trout, Stomach content; 2006)
Brazil (Guiana Dolphins, blubber; 2004-2005)
Brazil (Selected Dolphins, blubber; 1997-99)
Brazil (Franciscana dolphins, blubber 1997-99)
India (Ganges river dolphin, blubber; 1993-1996)
India - Bay of Bengal (Dolphins, blubber; 1997-99)
India (Irrawaddy dolphin, blubber 2000-01)
Russia (Baikai seal pups, blubber; 2000)
Canada - SW Ellesmere Is., Eureka (Female ringed seal, blubber; 1994)
Russia - Kalgalaksha Bay (Female ringed seal; blubber; 2001)
Russia - Kalgalaksha Bay (Male ringed seal; blubber; 2001)
USA - Florida Coast (Dolphins, blubber; 1991-2004)
Africa
Latin America
S-SE Asia
Developed region

Figure 2. Levels of PCBs from countries in the developed and developing regions: (a) levels in soil, ng/g dry weight; (b) levels in ambient air, ng/m^3; (c) levels in aquatic mammals, ng/g lipid weight. Year of sample collection is indicated in bracket. Central values ($ = AM; £ = 1/3 range; ¥ = GM; € = Median; * = single data point) are presented. The bar indicates minimum and maximum range if reported in the literature. S-SE Asia = South and Southeast Asia.

Possible evaporation of PBDEs under warm environment could be responsible for levels of PBDEs in the tropical areas where many developing countries are located [94].

Although there is a general lack of soil data of PBDEs for the general environment, PBDEs were monitored in soils around the vicinity of known point sources. For example, Eguchi et al. [95] identified crude recycling of e-waste as the main source of soil pollution by PBDEs in Asian developing countries after analyzing samples from 32 stations around e-waste disposal sites in India, Vietnam, Malaysia, Indonesia and Cambodia during 1999–2004. The contribution of leachates from landfills to the overall environmental burden is also highly significant due to the fact that most consumer products containing brominated flame retardants eventually end up in landfills [96]. Leachate samples collected from the same period from different landfills in South Africa showed wide variations in PBDE levels [97,98]. Levels of PBDEs in sediment samples collected from the sewer system of Hochiminh city in 2004 were found higher than samples collected from the estuary of Saigon-Dongnai River, suggesting a movement of PBDEs from urban sources to the surrounding aquatic environment [99]. PBDE levels in eggs of different bird species showed great variation in congener patterns; reflecting differences in trophic levels, migratory behavior, and distance to the source and exposure to different PBDE mixtures, with higher PBDE levels presented in species in close contact with human activities [100].

Levels of PBDEs in aquatic mammals from Cambodia [101], India [52,102], and Brazil [103,104] varied over two orders of magnitude (10–1,000 ng/g) (Figure 3b). Similar variation in PBDE levels was also observed for countries in the developed region such as East Greenland [105], United States [51,106], Russia [55], and Scotland - Isle of May [107]. Levels of PBDEs in aquatic mammals from McMurdo Sound–Antarctica were about one order of magnitude lower at about 1–2 ng/g [51] (Figure 3b). The highest PBDE levels in aquatic mammals were found in the Sao Paulo Coast of Brazil and San Francisco Bay in the United States. The Brazilian coast line, with an estimated population of about 11 million, is one of the most developed areas in the southern Atlantic. It has been impacted by several industries, tourism and agriculture which can be significant sources for PBDEs to the aquatic environment. In contrast, higher PBDE levels in the United States aquatic environment are mainly due to utilization of over 90% of PentaBDE produced globally [15]. PBDE congener patterns in aquatic mammals however, were influenced by the source of the contaminant, transport pathways, diet and species [108].

Although PBDE data for aquatic mammals in and around the African continent are lacking, high levels of the biogenic methoxylated PBDEs (MeO-BDEs) in Indo Pacific bottlenose (mean = 62,000 ng/g) and spinner dolphins (mean = 74,000 ng/g) from the Zanzibar channel in Tanzania were reported [59]. PBDEs were not measured in this study due to lack of available standards.

3. Levels of POPs in human blood in developing regions

Only a handful reports on POPs in human blood from the developing regions are available. Therefore, no summary figures are constructed in this review. More studies were carried out on human milk, which will subsequently be discussed. Since the analytical methods for measuring PCBs and DDTs in humans are very similar, DDTs and PCBs in human blood were often measured and reported together. Therefore, these two groups of POPs in human blood are discussed together in this section.

3.1. PCBs and DDTs

One study on maternal blood from women giving birth in South Africa showed that DDTs were detected in most samples and at higher levels (0.2–14,000 ng/g) than PCBs (0.27–20 ng/g) [109]. The results corroborate an earlier study in Nairobi, Kenya where DDTs (mean = 2,700 ng/g) were detected, but no PCBs, in maternal blood of mothers giving birth by caesarean section [110]. High DDE levels (mean = 380 μg/L) were also detected in blood among active pesticide handlers and harvesters in Ghana, Africa [111].

Human exposure to PCBs and DDTs can be influenced by many factors. Levels of DDTs in blood samples were lower in rural population (median DDE = 5 ng/mL) than in urban population (median DDE = 23 ng/mL) in Vietnam [112]. Mean Levels of DDTs in blood from Nagaon district in Northeast India had a positive correlation with age, showing 250 μg/L, 720 μg/L and 2,000 μg/L for age group < 25 year old, 25–50, and > 50, respectively [113]. The same study also found that DDTs were higher in males than in females. DDTs levels in blood were also linked to inhaled vapours of the pesticide [114,115] and diet [116,117]. Two time-trend studies suggested that levels of PCBs and DDTs are decreasing in human from West Africa [118]. Body mass index (BMI) showed no statistical association with levels of DDTs or PCBs in human serum in a Bolivian study [119], but demonstrated a positive association with DDTs in adipose tissue of women in Argentina [114].

3.2. PBDEs

Compared to PCBs and DDTs data, there is less data on PBDEs in human blood from the developing regions. However, the limited data indicated a positive association between PBDE levels in general population and urbanization [120,121]. For example, total PBDE blood levels in children living in industrial and urban areas of Mexico were approximately two times higher than those living in rural and municipal areas [122]. Proximity to waste disposal sites and consumption of fish were identified as two important exposure factors for PBDEs in children in Managua, Nicaragua [123]. In Asian developing countries, e-waste recycling plants and municipal solid waste dumping sites were also identified as major PBDE exposure sources for the general population [124]. In countries in the developed region, PBDE congeners in air and dust from e-waste recycling plants have been shown to have similar profiles as those in blood serum from workers in the plants [125-128]. In Africa, an increase in BDE-153 levels in the serum samples of police officers (n = 33) from Guinea Bissau during the study period (1990–2007) were observed [118]. This increase however, was baffling as there were no apparent exposure sources.

4. Comparison of levels of POPs in human milk (2000–present)

Unlike other matrices, human milk has been used as a viable sample type in a number of studies around the world. For the purpose of this review, we focused on studies in both developing and developed regions reported since year 2000 to provide a global perspective for the levels of POPs in human milk. Comparing levels of POPs amongst different studies to evaluate trends of specific environmental contaminants in human milk samples is a challenging task [129]. Data compatibility issue due to use of different analytical methods amongst various studies and inconsistency in the number of congeners being measured within a single group of contaminants such as PCBs or PBDEs are the two major problems. We have chosen only those congeners that were reported most consistently in human milk for comparing the levels among different studies. For PBDEs, these are BDE-47, BDE-99, BDE-100 and BDE-153; for PCBs, the congeners are PCB-138, PCB-153 and PCB-180. For DDTs, *p,p'*-DDT and *p,p'*-DDE were chosen. BDE-209 was excluded because it was not always reported in the PBDE studies despite the fact that it is has been reported in both the general population and in occupationally exposed individuals [130].

4.1. DDTs

DDTs in human milk from countries in the developing region were higher than the levels in the developed region as shown in Figure 4a [131-144]. This observation is consistent with findings from an earlier global trend study on DDTs more than a decade ago [145]. Just as it was observed more than a decade ago, the period that restrictions on the use and application of DDT were enforced in different regions correlated with the trends in the levels of DDTs observed in and between these regions [146]. Early restrictions and/or banning of DDT for agriculture in most parts of the developed regions has accounted for the gradual reductions of these compounds in human milk. DDT was banned in most of Western Europe in the late 60s and early 70s [147]. Although Mexico partially restricted the use of DDT in 1972 and banned its use in 1990, levels of DDTs in human milk sampled in 2000 and 2006 in Mexico were among the highest in the world (Figure 4a). Continuous use of DDT has been reported in both the northern and southern Vietnam, which may contribute to the high level of DDTs in human milk in Vietnam [134].

Ghana had the lowest level of DDTs among countries in the developing region. This should be interpreted as an isolated case since DDT is still used in Ghana to fight cotton pests, to control malaria and for artisanal fishing in the Ofin River [111].

It is important to mention that worldwide ban on DDT came into force in 2004. The United Nations convention provides an exemption for continued use of DDT for the purpose of disease-vector control in malaria-dense areas [148]. Highly infested mosquito areas in most parts of the developing regions have witnessed heavy applications of DDT, which almost immediately led to reductions in malaria cases, and consistently high levels of DDTs in human milk [149].

4.2. PCBs

Global distribution of PCBs in human milk is illustrated in Figure 4b [134,137,139,141,142,150-163].The levels in countries in the developed region (especially in Europe) seem to be quite uniform at around the 100 ng/g. On the other hand, data from the developing regions were scarce and showed huge variations. Available data from Mexico, Vietnam and Tunisia showed that levels of PCBs in these countries were comparable to those in Europe, while levels from Ghana and South Africa were much lower.

Similar PCB levels among countries in the developed region can be attributed to the regulations on the production and application of PCBs in the developed countries. The variation of PCB levels in countries in the developing region however shows a lack or ineffectiveness of such regulations. Lack of proper hazardous waste treatment facilities in the developing region coupled with mass import of obsolete electrical appliances such as transformers from developed countries lead to on-going human exposure to PCB in some of the developing countries. In Tunisia, for example, a 2004 report identified 1079 PCB contaminated transformers, representing 720 tons of liquid PCBs and 2900 tons of contaminated equipment in the country [163]. There are speculations that similar sites are dotted across countries not only in Africa but also in those located in S-SE Asian region. It is therefore necessary to investigate PCB pollution and all its potential sources, and to curb environmental exposures in these regions where no industrial production was reported [158].

4.3. PBDEs

PBDEs in human milk are shown in Figure 4c [136,148,152,157-172]. Levels of PBDEs in human milk from India, Vietnam and South Africa were about one to two orders of magnitude lower than those from other countries in the developing regions. Levels from other countries in the developing region were similar to those from European countries. Among the developed countries, the levels in Canada and the United States, depending on the studies, are higher than the levels in other developed countries especially those in Europe. The 2004 ban by the European Union on the production, use and import of penta-BDE and octa-BDE products has been credited for the drop in levels. The comparable levels in European countries, Japan and Australia and the developing regions should be interpreted with caution as the levels in the former are in a decreasing trend due to strict regulations previously alluded to, while those in the latter are on an upward trend. Samples collected in Ghana supports this upward trend in PBDE levels in human milk from 2004 to 2009 [158]. A similar upward trend of PBDE levels was observed in ambient air in another African country of Botswana (Figure 3c).

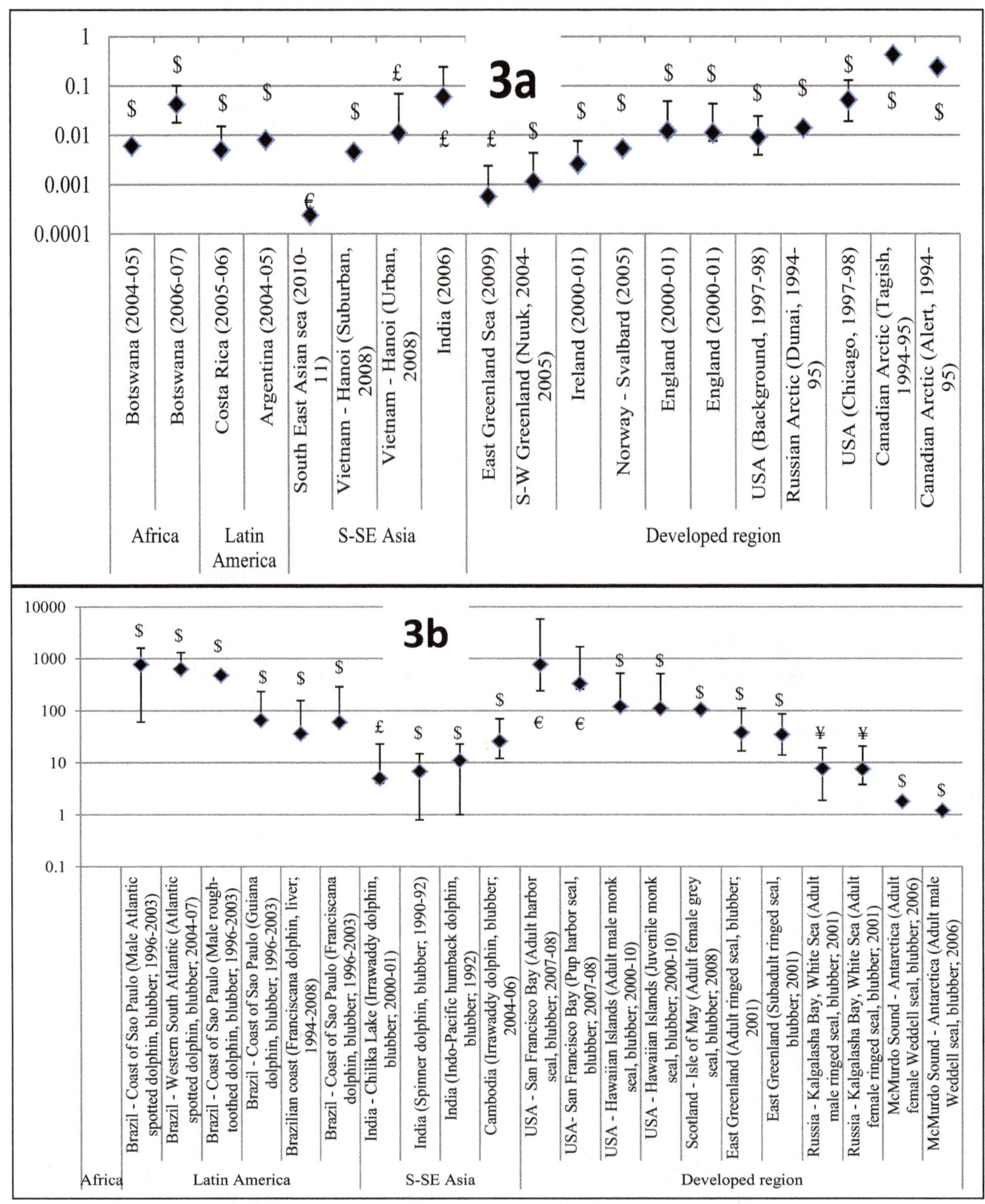

Figure 3. Levels of PBDEs from countries in the developed and developing regions: (a) levels in ambient air, ng/m^3; (b) levels in aquatic mammals, ng/g lipid weight. Year of sample collection is indicated in bracket. Central values ($ = AM; £ = 1/3 range; ¥ = GM; € = Median; * = single data point) are presented. The bar indicates minimum and maximum range if reported in the literature. S-SE Asia = South and Southeast Asia.

5. Conclusion and future directions

This review provided a global summary on three groups of POPs, namely DDTs, PCBs and PBDEs in the developing regions as well as some references to the levels in the developed regions. Among these three groups of POPs, PBDE levels in human milk are about an order of magnitude lower than PCBs and DDTs (Figure 4). One noticeable exception is Canada and United States where PBDE levels are comparable to DDTs and PCBs.

Despite of limited monitoring data in the developing regions, lower levels of PCBs and PBDEs, especially in aquatic mammals and in humans, are evident. Higher levels of DDTs in the developing regions, particularly in ambient air and in soil, might be as a result of historical and current use of DDT in agriculture and in disease control in some countries in the regions. Levels reported in the literature are inadequate to be representative of developing regions to establish spatial distribution of POPs in the regions. Data on temporal trends of POPs in the developing regions are particularly lacking. Some data included in this review, especially the levels of DDTs in soil from some countries in the developing region [30,33] are more than a decade old and so might not represent current levels.

PBDEs are relatively new POPs compared to DDTs and PCBs. Higher levels of PBDEs in countries in the developed region, particularly Canada and United States, could be the result of heavy use of PBDEs in the past two decades. Two separate studies measuring PBDEs in ambient air from Botswana [87,88] and in human milk from Ghana [158] showed a possible increase in PBDE levels in these countries. Such upward trend may serve as a warning sign that PBDE levels may on the rise in the developing regions. Therefore, continuing monitoring of PBDEs in the developing regions is crucial to detect possible rising PBDE pollution in these regions.

Most countries in the developed region have taken strong actions against the production and use of DDTs, PCBs and PBDEs as well as control of their emissions. Such effort is still weak in the developing regions due to several reasons of which financial could be considered paramount. For example, due to increased demand of food for domestic consumption and for export, fertilizers and pesticides are widely used in the developing regions [173]. Another concerning area is the export of obsolete electronics and electronic parts as secondhand electronic equipment, or for recycling, from countries in the developed region to some in the developing regions [174]. The latter practice will particularly impact the levels of PBDEs and other flame-retardants in developing regions.

Although this review have provided a global summary of POPs in the developing regions based on data currently available to us, we felt that more monitoring data are required. Several international monitoring activities such as Global Atmosphere Passive Sampling Network for measuring POPs in ambient air [175] and biomonitoring of POPs in human milk by the World Health Organization [176] will continue provide further information on the levels of POPs around the world including the developing regions. International monitoring activities also call for closer collaboration between countries in the developed region and those in the developing regions. Such collaboration is essential for the sharing of technologies and resources as well as generation of compatible data among countries for better interpretation on a global perspective [177].

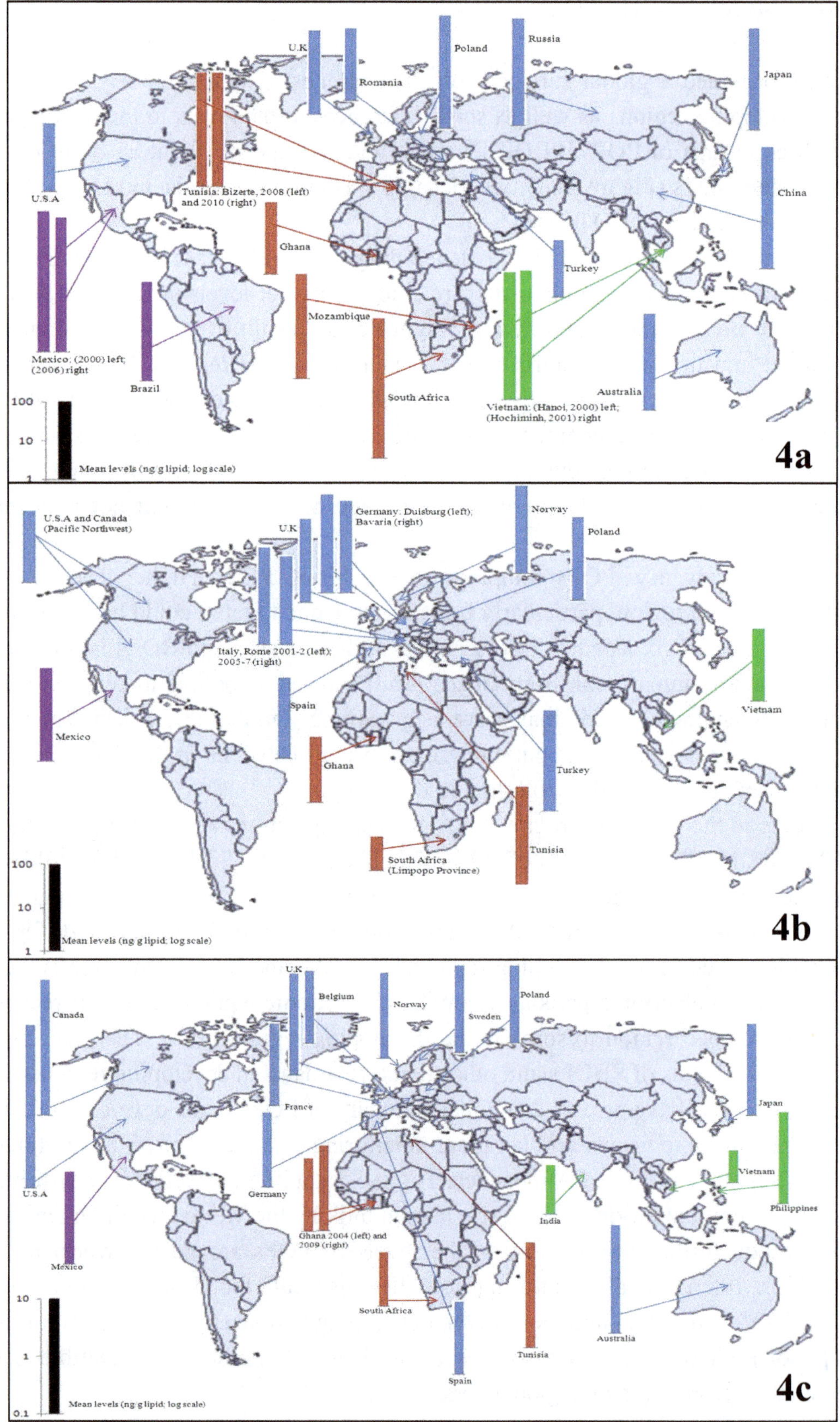

Figure 4. Arithmetic means of POPs in human milk (ng/g lipid weight) sampled post year 2000 from countries in the developed and developing regions: (a) DDTs; (b) PCBs; (c) PBDEs.

Acknowledgement

We thank many Health Canada colleagues, especially Mikin Patel and Mary Albert, for their review of the manuscript. P.M. thanks National Science and Engineering Council of Canada for the Government Laboratory Visiting Fellowship.

Conflict of interest

All authors declare no conflicts of interest in this paper.

References

1. Stockholm Convention, United Nations Environmental Program, 2011. Available from: http://chm.pops.int/default.aspx.
2. Carpenter DO (2006) Polychlorinated biphenyls (PCBs): routes of exposure and effects on human health. *Rev on Environ Health* 1: 1-23.
3. Longnecker MP, Gladen BC, Patterson DG Jr, et al. (2000) Polychlorinated biphenyl (PCB) exposure in relation to thyroid hormone levels in neonates. *Epidemiol* 11: 249-254.
4. Aoki Y (2001) Polychlorinated biphenyls, polychlorinated dibenzo-*p*-dioxin, polychlorinated dibenzo furans as endocrine disrupters – what we have learned from Yusho disease. *Environ Res* 86: 2-11.
5. White SS, Birnbaum LS (2009) An overview of the effects of dioxins and dioxin-like compounds on vertebrates, as documented in human and ecological epidemiology. *J Environ Sci Heal C* 27: 197-211.
6. Turusov V, Rakitsky V, Tomatis L (2002) Dichlorodiphenyltrichloroethane (DDT): ubiquity, persistence, and risks. *Environ Health Persp* 110: 125-128.
7. Longnecker MP, Klebanoff MA, Zhou H, et al. (2001) Association between maternal serum concentration of the DDT metabolite DDE and preterm and small-for-gestational-age babies at birth. *The Lancet* 358: 9276.
8. Alavanja MCR, Hoppin JA, Kamel F (2004) Health effects of chronic pesticide exposure: cancer and neurotoxicity. *Annu Rev Publ Health* 25: 155-197.
9. Rogan WJ, Chen A (2005) Health risks and benefits of bis(4-chlorophenyl)-1,1,1-trichloroethane (DDT). *The Lancet* 366: 763-773.
10. Nakata H, Kawazoe M, Arizono K, et al. (2002) Organochlorine pesticides and polychlorinated biphenyl residues in foodstuffs and human tissues from China: status of contamination, historical trend, and human dietary exposure. *Arch Environ Con Toxicol* 43: 473-480.
11. Nasreddine L, Parent-Massin D (2002) Food contamination by metals and pesticides in the European Union. Should we worry? *Toxicol Lett* 127: 29-41.
12. Jiang J-J, Lee C-L, Fang M-D, et al. (2011) Polybrominated biphenyl ethers and polychlorinated bipenyls in sediments of Southwest Taiwan: regional characteristics and potential sources. *Mar Pollut Bull* 62: 815-823.
13. Costa LG, Giordano G, Tagliaferri S, et al. (2008) Polybrominateddiphenyl ether (PBDE) flame retardants: environmental contamination, human body burden and potential adverse health effects. *Acta Biomed* 79: 172-183.

14. McDonald TA (2002) A Perspective on the potential health risks of PBDEs. *Chemosphere* 46: 745-755.
15. Yogui GT, Sericano JL (2009) Polybrominateddiphenyl ether flame retardants in the U.S. marine environment: a review. *Environ Int* 35: 655-666.
16. Aguilar A, Borrell A, Reijnders PJH (2002) Geographical and temporal variation in levels of organochlorine contaminants in marine mammals. *Mar Environ Res* 53: 425-452.
17. de Wit C, Alaee M, Muir D (2004) Brominated flame retardants in the arctic – an overview of spatial and temporal trends. *Organochlorine Compounds* 66: 3764-3769.
18. de Wit CA, Alaee M, Muir DCG (2006) Levels and trends of brominated flame retardants in the Arctic. *Chemosphere* 64: 209-233.
19. Lau MH, Leung KM, Wong SW, et al. (2012) Environmental policy, legislation and management of persistent organic pollutants (POPs) in China. *Environ Pollut* 165: 182-192.
20. Bao LJ, Maruya KA, Snyder SA, Zeng EY (2012) China's water pollution by persistent organic pollutants. *Environ Pollut* 163:100-108.
21. Moore SE, Huntington HP (2008) Arctic marine mammals and climate change: impacts and resilience. *Ecol Appl* 18: 157-165.
22. Burton A (2009) Toward DDT-free malaria control. *Environ Health Persp* 117: A334.
23. Manirakiza P,Akinbamijo O, Covaci A, et al. (2003) Assessment of organochlorine pesticide residues in West African City Farms: Banjul and Dakar case study. *Arch. Environ. Contam Toxicol* 44: 171-179.
24. Ngabe B, Bidleman BF (2006) DDT concentration in soils of Brazzaville, Congo. *B Environ Contam Toxicol* 76: 697-704.
25. Kishimba MA, Henry L, Mwevura H, et al. (2004)The status of pesticide pollution in Tanzania. *Talanta* 64: 48-53.
26. Ssebugere P (2008) Determination of persistent organic pollutant pesticides in soil in soil and fish from Kanungu District, Uganda. MSc Thesis, Department of Chemistry, Makerere University, Kampala, Uganda.
27. Ssebugere P, Wasswa J, Mbabazi J, et al. (2010) Organochlorine pesticides in soils from south-western Uganda. *Chemosphere* 78: 1250-1255.
28. Westbom R, Hussen A, Megersa N, et al. (2008) Assessment of organochlorine pesticide pollution in Upper Awash Ethiopian state farm soils using selective pressurized liquid extraction. *Chemosphere* 72: 1181-1187.
29. Torres JPM, Pfeiffer WC,Markowitz S, et al. (2002) Dichlorodiphenyltrichloroethane in Soil, River Sediment, and Fish in the Amazon in Brazil. *Environ Res* 88:134-139.
30. Urzua H, Romerot J, Ruizt VM (1986) Effect of p,p' DDT on nitrogen fixation of white clover in volcanic soils of Chile. *MIRCEN J* 2: 365-372.
31. Yanez L, Ortiz-Pérez D, Lilia E, et al. (2002) Levels of dichlorodiphenyltrichloroethane and deltamethrin in humans and environmental samples in malarious areas of Mexico. *Environ Res* 88: 174-181.
32. Toan VU, Thao VU, Walder J, et al. (2007) Contamination by selected organochlorine pesticides (OCPs) in surface soils of Hanoi, Vietnam. *B Environ Contam Toxicol* 78: 195-200.
33. Pillai MKK (1986) Pesticide pollution of soil, water and air in Delhi area, India. *Sci Total Environ* 55: 321-327.

34. Wong F, Robson M, Diamond ML, et al. (2009) Concentrations and chiral signatures of POPs in soils and sediments:A comparative urban versus rural study in Canada and UK. *Chemosphere* 74: 404-411.
35. Aigner EJ, Leone AD, Falconer RL (1998) Concentrations and enantiometric ratios of organochlorine pesticides in soils from the U.S Corn Belt. *Environ Sci Technol* 32: 1162-1168.
36. Wong F, Harner T, Liu QT, Diamond ML (2004) Using experimentaland forestsoils to investigate the uptake of polycyclicaromatichydrocarbons(PAHs)along an urban–rural gradient. *Environ Pollut* 129: 387-398.
37. Bidleman TF, Leone AD (2004) Soil-air exchange of organochlorine pesticides in the Southern United States. *Environ Pollut* 128:49-57.
38. Kannan K, Battula S, Loganathan BG, et al. (2003) Trace organic contaminants, including toxaphene and trifluralin, in cotton field soils from Georgia and South Carolina, USA. *Arch Environ Contam Toxicol* 45:30-6.
39. Ngabe B, Bidleman BF (1992) Occurrence and vapour particle partitioning of heavy particle compounds in ambient air in Brazzaville, Congo. *Environ Pollut* 76: 147-156.
40. Karlsson H, Muir DCG, Teixiera CF, et al. (2000) Presistent chlorinated pesticides in air, water, and precipitation from the Lake Malawi area, Southern Africa. *Environ Sci Tech* 34: 4490-4495.
41. Larsson P, Berglund O, Backe C, et al. (1995) DDT-Fate in tropical and temperate regions. *Naturwissenschaften* 82: 559-561.
42. Alegria HA, Wong F, Jantunen LM,et al. (2008) Organochlorine pesticides and PCBs in air of southern Mexico (2002-2004). *Atmos Environ* 42: 8810-8818.
43. Meire RO, Lee SC, Yao Y,et al. (2012) Seasonal and altitudinal variations of legacy and current use pesticides in the Brazilian tropical and subtropical mountains. *AtmosEnviron* 59: 108-116.
44. Hong SH, Yim UH, Shim WJ, et al. (2008) Persistent organic chlorine residues in estuarine and marine sediments from Ha Long Bay, HaiPhong Bay and Ba Lat Estuary, Vietnam. *Chemosphere*72: 1193-1202.
45. Iwata H, Tanabe S, Sakai N, et al. (1994) Geographical distribution of persistent organochlorines in air, water and sediments from Asia and Oceania, and their implications for global redistribution from lower latitudes. *Environ Pollut* 85: 15-33.
46. Daly GL, Lei YD, Teixeira C,et al.(2007) Pesticides in Western Canadian mountain air and soil. *Environ Sci Technol* 41: 6020-6025.
47. Hoff RM, Muir DCG, Grift NP (1992) Annual cycle of polychlorinated biphenyls and organohalohen pesticides in air in Southern Ontario. 1. Air concentration data. *Environ Sci Technol* 26: 266-275.
48. Ondarza PM, Gonzalez M, Fillmann G, Miglioranza KSB (2011) Polybrominateddiphenyl ethers and organochlorine compound levels in brown trout (*Salmotrutta*) from Andean Patagonia, Argentina. *Chemosphere* 83: 1597-1602.
49. Lailson-Brito J, Dorneles PR, Azevedo-Silver CE, et al. (2012) Organochlorine compound accumulation in delphinids from Rio de Janeiro State, southeastern Brazilian coast. *Sci Total Environ* 433: 123-131.
50. Hobbs KE, Muir DCG, Born EW, et al. (2003) Levels and pattern of persistent organochlorine in minke whale (*Balaenopteraacutorostrata*) stocks from the North Atlantic and European Arctic. *Environ Pollut* 121: 239-252.

51. Lopez J, Boyd D, Ylitalo GM, et al. (2012) Persistent organic pollutants in the endangered Hawaiian monk seal (*Monachusschauinsland*) from the main Hawaiian Islands. *Mar Pollut Bull* 64: 2558-2598.
52. Kannan K, Ramu K, Kajiwara N, et al. (2005) Organochlorine pesticides, polychlorinated biphenyls, and polybrominateddiphenyl ethers in Irrawaddy Dolphins from India. *Arch Environ Contam Toxicol* 49: 415-420.
53. Senthilkumar K, Kannan K, Sinha RK, et al. (1999) Bioaccumulation of polychlorinated biphenyl congeners and organochlorine pesticides in Ganges River dolphins. *Environ Toxicol Chem* 18: 1511-1520.
54. Trumble SJ, Robinson EM, Noren SR, et al. (2012) Assessment of legacy and emerging persistent organic pollutants in Weddell seal tissue (*Leptonychotesweddellii*) near McMurdo Sound, Antarctica. *Sci Total Environ* 439: 275-283.
55. Muir D, Savinova T, Savinova V, et al. (2003) Bioaccumulation of PCBs and chlorinated pesticides in seals, fishes and invertebrates from the White Sea, Russia. *Sci Total Environ* 306: 111-131.
56. Savinov V, Muir DCG, Svetochev V, et al. (2011) Persistent organic pollutants in ringed seals from the Russian Arctic.*Sci Total Environ* 409: 2734-2745.
57. Schlenk D, Sapozhnikova Y, Cliff G (2005) Incidence of organochlorine pesticides in muscle and liver tissues of South African great white sharks *Carcharodoncarcharias*. *Mar Pollut Bull* 50: 208-211.
58. Vetter W, Weichbrodt M, Scholz E, et al. (1999) Levels of organochlorines (DDT, PCBs, toxaphene, chlordane, dieldrin, and HCHs) in blubber of South African fur seals (*Arctocephaluspusilluspusillus*) from Cape Cross/Namibia. *Mar Pollut Bull* 38: 830-836.
59. Mwevura H, Amir OA, Kishimba M, et al. (2010)Organochlorine compounds in blubber of Indo-Pacific bottlenose dolphin (*Tursiopsaduncus*) and spinner dolphin (*Stenellalongirostris*) from Zanzibar, Tanzania. *Environ Pollut* 158: 2200-2207.
60. Borrell A, Cantos G, Aguilar A, et al. (2007) Concentrations and patterns of organochlorine pesticides and PCBs in Mediterranean monk seals (*Monachusmonachus*) from Western Sahara and Greece. *Sci Total Environ* 381: 316-325.
61. Aguilar A, Borrell A, Pastor T (1999) Biological factors affecting variability of persistent pollutant levels in cetaceans. *J Cetacean Res Manage* 1: 83-116.
62. Batterman S,Chernyak S, Yoganathan J, et al. (2008) PCBs in air, soil and milk in industrialized and urban areas of KwaZulu-Natal, South Africa. *Environ Pollut* 157: 654-663.
63. Quinn L, Pieters R, Nieuwoudt C, et al. (2009) Distribution profiles of selected organic pollutants in soils and sediments of industrial, residential and agricultural areas of South Africa. *J Environ Monit* 11: 1647-1657.
64. Barra R, Popp P, Quiroz R, et al. (2005) Persistent toxic substances in soils and waters along an altitudinal gradient in the Laja River Basin, Central Southern Chile. *Chemosphere* 58: 905-915.
65. Rissato SR., Galhiane MS, Ximenes VF, et al. (2006) Organochlorine pesticides and polychlorinated biphenyls in soil and water samples in the north eastern part of Sao Paulo State, Brazil. *Chemosphere* 65: 1949-1958.
66. Thao VD, Kawano M, Matsuda M, et al. (1993) Chlorinated hydrocarbon insecticide and polychlorinated biphenyl residues in soils from southern provinces of Vietnam. *Int J Environ An Ch* 50: 147-159.

67. Toan VU, Thao VU, Walder J (2007) Level and distribution of polychlorinated biphenyls (PCBs) in surface soils from Hanoi, Vietnam. *Bull Environ Contam Toxicol* 78: 211-216.
68. Li Y-F, Harner T, Liu L, et al. (2010) Polychlorinated Biphenyls in Global Air and Surface Soil: Distributions, Air–Soil Exchange, and Fractionation Effect. *Environ Sci Tech* 44: 2784-2790.
69. Breivik K, Sweetman A, Pacyna JM, Jones KC (2002) Towards a global historical emission inventory for selected PCB congeners – a mass balance approach 1. Global production and consumption. *Sci Total Environ* 290: 181-198.
70. Gioia R, Eckhardt S, Breivik K, et al. (2011) Evidence for major emissions of PCBs in the West African region. *Environ Sci Tech* 45: 1349-1355.
71. Pozo K, Harner T, Shoeib M, et al. (2004) Passive-sampler derived air concentrations of persistent organic pollutants on a North-South Transect in Chile. *Environ SciTechnol* 38: 6529-6537.
72. Gasic B, Moeckel C, MacLeod M, et al. (2009) Measuring and modeling short-term variabilityof PCBs in air and characterization of urban source strength in Zurich, Switzerland. *Environ SciTechnol* 43: 769-776.
73. Halsall CJ, Lee RGM, Burnett V, et al. (1995) PCBs in UK urban air. *Environ SciT echnol* 29: 2368-2376.
74. Kim K-S, Masunaga S (2005) Behavior and source characteristic of PCBS in urban ambient air of Yokohama, Japan. *EnvironPollut* 138: 290-298.
75. Mari M, Schuhmacher M, Feliubadalo J, Domingo JL (2008) Air concentrations of PCDD/Fs, PCBs and PCNs using active and passive air samplers. *Chemosphere* 70: 1637-1643.
76. Stranberg B, Dodder NA, Basu I,Hites RA (2001) Concentrations and spatial variations of polybrominateddiphenyl ethers and other organohalogen compounds in Great Lakes air. *Environ Sci Technol* 35: 1078-1083.
77. Buehler SS, Hites RA (2002) It's in the air: A look at the Great Lakes Integrated Atmospheric Deposition Network, *Environ Sci Technol* 36: 354A-359A.
78. Shen L, Wania F, Lei YD, et al. (2006) Polychlorinated biphenyls and polybrominateddiphenyl ethers in the North American atmosphere. *Environ Pollut* 144: 434-444.
79. Eckhardt K, Breivik K, Mano S, Stohl A (2007) Record high peaks in PCB concentrations in the Arctic atmosphere due to long-range transportof biomass burning emissions. *Atmos Chem Phys* 7: 4527–4536.
80. Kajiwara N, Matsuoka S, Iwata H, et al. (2004) Contamination by persistent organochlorines in Cetaceans incidentally caught along Brazilian Coastal Waters. *Arch Environ Contam Toxicol* 46: 124–134.
81. Alonso MB, Marigo J, Bertozzi CP, et al. (2010) Occurrence of chlorinated pesticides and polychlorinated biphenyls (PCBs) in Guiana Dolphines(*ScotaliaGuianensis*) from Urbatuba and BaixadaSantista, São Paulo, Brazil. *Lat Am J Aquat Mammal* 8: 123-130.
82. Tomy GT, Muir DCG, Stern GA, Westmore JB (2000) Levels of C_{10}-C_{13}polychloro-*n*-alkanes in marine mammals from the arctic and the St. Lawrence River estuary. *Environ Sci Technol* 34: 1615-1619.
83. Karuppiah S, Subramanian A, Obbard JP (2005) Organochlorine residues in odontocete species from the southeast coast of India. *Chemosphere* 60: 891-897.

84. Tsydenova O, Minh TB, Kajiwara N, et al. (2004) Recent contamination by persistent organochlorines in Baikal seal (*Phocasibirica*) from Lake Baikal, Russia. *Mar Pollut Bull* 48: 749-758.
85. Johnson-Restrepo B, Kannan K,Addink R, Adams DH (2005) Polybrominateddiphenyl ethers and polychlorinated biphenyls in a marine food web of Coastal Florida. *Environ SciTechnol* 39: 8243-8250.
86. Minh HN, Someya M, Minh TB, et al. (2004) Persistent Organochlorine residues in human breast milk from Hanoi and Hochiminh city, Vietnam: contamination, accumulation kinetics and risk assessment for infants. *EnvironPollut* 28: 431-441.
87. Pozo K, Harner T, Wania F, et al. (2006) Towards a Global Network for Persistent Organic Pollutants in Air: Results from the GAPS Study. *Environ Sci Technol* 40: 4867–4873.
88. Shunthirasingham C, Gouin T, Lei YD, et al. (2010) Year-round measurements of PBDEs in the atmosphere of tropical Costa Rica and subtropical Botswana. 5th International Symposium on Brominated Flame Retardants, April 7-9, 2010, Kyoto, Japan.
89. Zhang G,Chakraborty P, Li J, et al. (2008) Passive atmospheric sampling of organochlorine pesticides, polychlorinated biphenyls, and polybrominateddiphenyl ethers in urban, rural, and wetland sites along the coastal length of India. *Environ SciTechnol* 15: 8218-8223.
90. Möller A, Xie Z, Cai M, et al. (2012) Brominated flame retardants and dechlorane plus in the marine atmosphere from Southeast Asia toward Antarctica. *Environ SciTechnol*46: 3141-3148.
91. Lee RGM, Thomas GO, Jones KC (2004) PBDEs in the atmosphere of three locations in Western Europe. *Environ Sci Technol* 38: 699-706.
92. Möller A, Xie Z, Sturm R, Ebinghaus R (2011) Polybrominateddiphenyl ethers (PBDEs) and alternative brominated flame retardants in air and seawater of the European Arctic. *Environ Pollut* 159: 1577-1583.
93. Bossi R, Skov H, Vorkamp K, et al. (2008) Atmospheric concentrations of organochlorinepesticides, polybrominateddiphenyl ethers and polychloronaphthalenes in Nuuk, South-West Greenland. *Atmos Environ* 42: 7293-7303.
94. Tanabe S (2002) Contamination and toxic effects of persistent endocrine disrupters in marine mammals and birds. *Mar Pollut Bull* 45: 69-77.
95. Eguchi A, Isobe T, Subramanian A, et al. (2009) Contamination by brominated flame retardants in soil samples from open dumping sites of Asian developing countries. In: Interdisciplinary Studies on Environmental Chemistry-Environmental Research in Asia. Eds.: Obayashi Y., Isobe T., Subramanian A., Suzuki S and Tanabe S. *TERRAPUB* 143-151.
96. Daso AP, Fatoki OS, Odendaal JP, Okonkwo JO (2010) A review on sources of brominated flame retardants and routes of human exposure with emphasis on polybrominateddiphenyl ethers. *Environ Rev* 18: 239-254.
97. Daso AP, Fatoki OS, Odendaal JP, Olujimi OO (2013) Polybrominateddiphenyl ethers (PBDEs) and 2,2’,4,4’,5,5’-hexabromobiphenyl (BB-153) in landfill leachate in Cape Town, South Africa. *Environ Monit Assess* 185: 431-439.
98. Odusanya, DO, Okonkwo JO, Botha B (2009) Polybrominateddiphenyl ethers (PBDEs) in leachates from selected landfill sites in South Africa. *Waste Manag* 29: 96-102.
99. Minh NH, Minh TB, Isobe T, Tanabe S (2010) Contamination of polybromodiphenyl ethers (PBDEs) in sewer system of Hochiminh City and Estuary of Saigon-Dongnai River. BFR 2010.

100. Quinn LP (2010) Assessment of organic pollutants in selected wild and domesticated bird eggs from Gauteng, South Africa. PhD thesis, School of Environmental Sciences and Development 2010, Potchefstroom Campus of the North-West University.
101. Dove V (2009) Mortality investigation of the Mekong Irrawaddy River Dolphin (*Orcaellabrevirostris*) in Cambodia based necropsy sample analysis. WWF Technical Report 2009. Available from: http://awsassets.panda.org/downloads/necropsy_report_irrawaddy_mortality_final.pdf.
102. Kajiwara N, Kamikawa S, Ramu K, et al. (2006) Geographical distribution of polybrominateddiphenyl ethers (PBDEs) and organochlorines in small cetaceans from Asian waters. *Chemsophere* 64: 287-295.
103. de la Torre A, Alonso MB, Martinez MA, et al. (2012) Dechlorane-related compounds in Franciscana dolphin (*Pontoporiablainvillei*) from Southeastern and Southern Coast of Brazil. *Environ SciTechnol* 46: 12364-12372.
104. Yogui GT, Santos MCO, Bertozzi CP, et al. (2011) PBDEs in the blubber of marine mammals from coastal areas of Sao Paulo, Brazil, southwestern Atlantic. *Mar Pollut Bull* 62: 2666-2670.
105. Vorkamp K, Christensen JH, Riget F (2004) Polybrominateddiphenyl ethers and organochlorine compounds in biota from the marine environment of East Greenland. *Sci of the Total Environ* 331: 143-155.
106. Klosterhaus SL, Stapleton HM, La Guardia MJ, et al. (2012) Brominated and chlorinated flame retardants in San Francisco Bay sediments. *Environ Inter* 47: 56-65.
107. Leonel J, Taniguichi S, Sasaki DK, et al. (2012) Contamination by chlorinated pesticides, PCBs and PBDEs in Atlantic spotted dolphin (*Stenellafrontalis*) in western South Atlantic. *Chemosphere* 86: 741-746.
108. Berghe MV, Weijs L, Habran S, et al. (2012) Selective transfer of persistent organic pollutants and their metabolites in grey seals during lactation. *Environ Inter* 46: 6-15.
109. Rollin HB, Sandanger TM, Hansen L, et al. (2009) Concentration of selected persistent organic pollutants in blood from delivering women in South Africa. *Sci Total Environ* 408: 146-152.
110. Kanja LW, Skaare JU, Ojwang SBO, et al. (1992) A comparison of organochlorine pesticide residues in maternal adipose tissue, maternal blood, cord blood and human milk from mother/infant pairs. *Ach. Environ Contam Toxcol* 22: 21-24.
111. Ntow WJ (2001) Organochlorine pesticides in water, sediment crops and human fluids in a farming community in Ghana. *Arch Environ ContamToxicol* 40: 557-563.
112. Schecter A, Toniolo P, Dai LC, et al. (1997) Blood levels of DDT and breast cancer risk among women living in the north of Vietnam.*Arch Environ ContamToxicol* 33: 453-456.
113. Mishra K, Sharma RC, Kumar S (2011) Organochlorine pollutants in human blood and their relation with age, gender and habitat from North-east India. *Chemosphere* 85: 454-464.
114. Munoz-de-Toro M, Beldomenica HR, Garcia SR, et al. (2006) Organochlorine levels in adipose tissue of women from a littoral region of Argentina. *Environ Research* 102: 107-112.
115. Wong F, Alegria HA, Jantunen LM, et al. (2008) Organochlorine pesticides in soil and air of southern Mexico: chemical profile and potential for soil emissions. *Atmos Environ* 42: 7737-7745.
116. Waliszewski SM, Villalobos-Pietrini R, Gómez-Arroyo S, et al. (2003) Persistent organochlorine pesticides in Mexican butter. *Food Addit Contam* 20: 361-367.

117. Waliszewski SM, Villalobos-Pietrini R, Gómez-Arroyo S, et al. (2003) Persistent organochlorine pesticide levels in cow's milk samples from tropical regions of Mexico. *Food Addit Contam* 20: 270-275.
118. Linderholm L, Biague A, Mansson F, et al. (2010) Human exposure to persistent organic pollutants in West Africa – a temporal trend study from Guinea-Bissau. *Environ Inter* 36: 675-682.
119. Arrebola JP, Cuellar M, Claure E, et al. (2012) Concentrations of Organochlorine pesticides and polychlorinated biphenyls in human serum and adipose tissue from Bolivia. *Environ Res* 112: 40-47.
120. Lopez D, Athanasiadou D, Athanassiadis I, et al. (2006) Estudio preliminar sobre los niveles de exposicion a PBDEs en sangre y leche materna en Mexico. *ActaToxicol Argent*14: 52-54.
121. Orta-Garcia ST, Leon-Moreno LC, Gonzalez-Vega C, et al. (2012) Assessment of the Levels of PolybrominatedDiphenyl Ethers in Blood Samples from Guadalajara, Jalisco, Mexico. *Bull Environ Contam Toxicol* 89, 925–929.
122. Pérez-Maldonado IN, Ramírez-Jiménez MR, Martínez-Arévalo LP, et al. (2009) Exposure assessment of polybrominateddiphenyl ethers (PBDEs) in Mexican children. *Chemosphere* 75: 1225-1220.
123. Athanasiadou M, Cuadra SN, Marsh G, et al. (2008) Polybrominateddiphenyl ethers (PBDEs) and bioaccumulativehydroxylated PBDE metabolites in young humans from Managua, Nicaragua. *Environ Health Persp* 116: 400-408.
124. Muto M,Isobe T, Ramu K, et al. (2012) Contamination of brominated flame retardants in human hair from e-waste recycling site in Vietnam. Interdisciplinary studies on environmental chemistry-Environmental pollution and ecotoxicology. Eds.: Kawaguchi, M.; Misaki, K.; Sato, H.; Yokokawa, T.; Itai, T.; Nguyen, T. M.; Ono, J.; Tanabe, S. *TERRAPUB* 229-237.
125. Jakobsson K, Thuresson K, Rylander L, et al. (2002) Exposure to polybrominateddiphenyl ethers and tetrabromobisphenol A among computer technicians. *Chemosphere* 46: 709-716.
126. Sjodin A, Hagmar L, Klasson-Wehler E, et al. (1999) Flame retardant exposure: polybrominateddiphenyl ethers in blood from Swedish workers. *Environ Health Persp* 107: 643-648.
127. Sjodin A,Carlsson H, Thuresson K, et al. (2001) Flame retardants in indoor air at an electronics recycling plant and at other work environments. *Environ Sci Technol* 35: 448-454.
128. Thuresson K, Jakobsson K, Hagmar L, et al. (2002) Work related exposure to brominated flame retardants when recycling metals from printed circuit boards. *Organohalogen Compounds* 58: 249.
129. Alivernini S, Battistelli CL, Turrio-Baldassarri L (2011) Human milk as a vector and as an indicator of exposure to PCBs and PBDEs: temporal trend of samples collected in Rome. *Bull Environ Contam Toxicol* 87: 21–25.
130. Sjodin A, Patterson DG Jr, Bergman A (2003) A review on human exposure to brominated flame retardants--particularly polybrominateddiphenyl ethers. *Environ Inter* 29: 829-839.
131. Azeredo A, Torres JPM, Fonseca MF, et al. (2008) DDT and its metabolites in breast milk from the Madeira River basin in the Amazon, Brazil. *Chemosphere* 73: S246–S251.
132. Cok I, Donmez MK, Karakaya AE (2004) Levels and trends of chlorinated pesticides in human breast milk from Ankara residents: comparison of concentration in 1984 and 2002. *Bull. Environ Contam Toxicol* 72: 522-529.

133. Ennaceur S, Gandoura N, Driss MR (2008) Distribution of polychlorinated biphenyls and organochlorine pesticides in human breast milk from various locations in Tunisia: Levels of contamination, influencing factors, and infant risk assessment. *Environ Res* 108: 86-93.
134. Haraguchi K, Koizumi A, Inoue K, et al. (2009) Levels and regional trends of persistent organochlorines and polybrominateddiphenyl ethers in Asian breast milk demonstrate POPs signatures unique to individual countries. *Environ Inter* 35: 1072-1079.
135. Hernik A,Góralczyk K, Struciński P, et al. (2011) Polybrominateddiphenyl ethers, polychlorinated biphenyls and Organochlorine pesticides in human milk as markers of environmental exposure to these compounds. Ann. Agric. *Environ Med* 18: 113-118.
136. Johnson-Restrepo B, Addink R, Wong C, et al. (2007) Polybrominateddiphenyl ethers and organochlorine pesticides in human breast milk from Massachusetts, USA. *J Environ Monit* 9: 1205–1212.
137. Kalantzi OI, Martin FL, Thomas GO, et al. (2004) Different levels of polybrominateddiphenyl ethers (PBDEs) and chlorinated compounds in breast milk from two U.K regions. *Environ Health Persp* 112: 1085-1091.
138. Kunisue T, Someya M, Kayama F, et al. (2004) Persistent organochlorines in human breast milk collected from primiparae in Dalian and Shenyang, China. *EnvironPollut* 131: 381-392.
139. Manaca MN, Grimalt JO, Sunyer J, et al. (2011) Concentration of DDT compounds in breast milk from African women (Manhica, Mozambique) at the early stages of domestic indoor spraying with this insecticide. *Chemosphere* 85: 307-314.
140. Mueller JF, Harden F, Toms L-M, et al. (2008) Persistent Organochlorine pesticides in human milk samples from Australia. *Chemosphere* 70: 712-720.
141. Ntow WJ, Tagoe LM, Drechsel P,et al. (2008) Accumulation of Organochlorine contaminants in milk and serum of farmers from Ghana. *Environ Res* 106: 17-26.
142. Rodas-Ortiz JP, Ceja-Moreno V, Gonzalez-Navarrete RL, et al. (2008) Organochlorine Pesticides and Polychlorinated Biphenyls Levels in Human Milk from Chelem, Yucatan, Mexico. *Bull Environ Contam Toxicol* 80: 255-259.
143. Sereda S (2005) Pyrethroid and DDT residues in human breast milk from KwaZulu Natal, South Africa. *Epidemiology*16: S157.
144. Tsydenova OV, Sudaryanto A, Kajiwara N, et al. (2007) Organohalogen compounds in human breast milk from Republic of Buryatia, Russia. *Environ Pollut* 146: 225-232.
145. Smith D (1999) Worldwide trends in DDT levels in human breast milk. *Epidemiol* 28: 179-188.
146. Polder A, Thomsen C, Lindström G, et al. (2008) Levels and temporal trends of chlorinated pesticides, polychlorinated biphenyls and brominated flame retardants in individual human breast milk samples from northern and southern Norway. *Chemosphere* 73: 14-23.
147. Solomon GM, Weiss PM (2002) Chemical contaminants in breast milk: time trends and regional variability. *Environ Health Persp* 110: A339-347.
148. Darnerud PO, Aune M, Larsson L, et al. (2011) Levels of brominated flame retardants and other persistent organic pollutants in breast milk samples from Limpopo province, South Africa. *Sci of the Total Environ.* 19: 4048-4053.
149. Bouwman H, Sereda B, Meinhardt HM (2006). Simultaneous presence of DDT and pyrethroid residues in human breast milk from a malaria endemic area in South Africa. *Environ Pollut* 144: 902-917.

150. Cok I, Gorucu E, Satıroglu HM, et al. (2003) Polychlorinated biphenyls (PCBs) levels in human breast milk samples from Turkish mothers. *Bull Environ Contam Toxicol* 70: 41-45.

151. Costopoulou D, Vassiliadou I, Papadopoulos A, et al. (2006) Levels of dioxins, furans and PCBs in human serum and milk of people living in Greece. *Chemosphere* 65: 1462-1469.

152. Gomara B, Herrero L, Pacepavicius G, et al. (2011) Occurrence of co-planer polybrominated/chlorinated biphenyls (PXBs), polybrominateddiphenyl ethers (PBDEs) and polychlorinated biphenyls (PCBs) in breast milk of women from Spain. *Chemosphere* 83: 799-805.

153. IngelidoAM, Ballard T, Dellatte E, et al. (2007) Polychlorinated biphenyls (PCBs) and polybrominateddiphenyl ethers (PBDEs) in milk from Italian women living in Rome and Venice. *Chemosphere* 67: S301–S306.

154. Raab U, Preiss U, Albrecht M, et al. (2008) Concentrations of polybrominateddiphenyl ethers, organochlorine compounds and nitro musks in mother's milk from Germany (Bavaria). *Chemosphere* 72: 87-94.

155. She J, Holden A, Sharp M, et al. (2007) Polybrominateddiphenyl ethers (PBDEs) and polychlorinated biphenyls (PCBs) in breast milk from the Pacific Northwest. *Chemosphere* 67: S307-S317.

156. Wittsiepe J, Furst P, Schrey P, et al. (2007) PCDD/F and dioxin-like PCB in human blood and milk from German mothers. *Chemosphere* 67: S286-S294.

157. Antignac J-P, Cariou R, Zalko D, et al. (2009) Exposure assessment of French women and their newborn to brominated flame retardants: determination of tri- to deca- polybromodiphenyl ethers (PBDEs) in maternal adipose tissues, serum, breast milk and cord serum. *Environ Pollut* 157: 164-173.

158. Asante KA, Kumi SA, Nakahiro K, et al. (2011) Human exposure to PCBs, PBDEs and HBCDs in Ghana: temporal variation, sources of exposure and estimation of daily intakes by infants. *Environ Inter* 37: 921-928.

159. Colles A, Koppen G, Hanot V, et al. (2008) Fourth WHO-coordinated survey of human milk for persistent organic pollutants (POPs): Belgian results. *Chemosphere* 73: 907–914.

160. Devanathan G, Isobe T, Subramanian A, et al. (2012) Contamination status of polychlorinated biphenyls and brominated flame retardants in environmental and biota samples from India. Interdisciplinary Studies on Environmental Chemistry-Environmental Pollution and Ecotoxicology. Eds.: Kawaguchi M., Misaki K., Sato H., Yokokawa T., Itai T., Nguyen T.M., Ono J and Tanabe S. *TERRAPUB* 269-277.

161. Eslami B, Koizumi A, Ohta S, et al. (2006) Large-scale evaluation of the current level of polybrominateddiphenyl ethers (PBDEs) in breast milk from 13 regions of Japan. *Chemosphere* 63: 554–561.

162. Furst P (2006) Dioxins, polychlorinated biphenyls and other organohalogen compounds in human milk: Levels, correlations, trends and exposure through breastfeeding. *MolNutr Food Res.* 50: 922 – 933.

163. Hassine SB, Ameur WB, Gandoura N, Driss RM (2012) Determination of chlorinated pesticides, polychlorinated biphenyls, and polybrominateddiphenyl ethers in human milk from Bizerte (Tunisia) in 2010. *Chemosphere* 89: 369–377.

164. Jaraczewska K, Lulek J, Covaci A, et al. (2006) Distribution of polychlorinated biphenyls, organochlorine pesticides and polybrominateddiphenyl ethers in human umbilical cord serum, maternal serum and milk from Wielkopolska region, Poland. *Sci Total Environ* 372: 20–31
165. Lignell S, Aune M, Darnerud PO, et al. (2011) Large variation in breast milk levels of organohalogenated compounds is dependent on mother's age, changes in body composition and exposures early in life. *J Environ Monit* 13: 1607-1616.
166. Lopez D, Athanasiadou M, Athanassiadis I, et al. (2010) A Preliminary Study on PBDEs and HBCDD in Blood and Milk fromMexican Women. 5th International Symposium on Brominated Flame Retardants, April 7-9, 2010, Kyoto, Japan.
167. Malarvannan G, Kunisue T, Isobe T, et al. (2009) Organohalogen compounds in human breast milk from mothers living in Payatas and Malate, the Philippines: Levels, accumulation kinetics and infant health risk. *Environ Pollut* 157: 1924 - 1932.
168. Schecter A, Pavuk M, Päpke O, et al. (2003) PolybrominatedDiphenyl Ethers (PBDEs) in U.S. Mothers' Milk. *Environ Health Persp* 111: 1723-1729.
169. Siddique S, Xian Q, Abdelouahab N, et al. (2011) Levels of dechlorane plus and polybrominateddiphenyl ethers in human milk in two Canadian cities. *Environ Inter* 39: 50-55.
170. Thomsen C, Stigum H, Frøshaug M, et al. (2010) Determinants of brominated flame retardants in breast milk from a large scale Norwegian study. *Environ Inter* 36: 68-74.
171. Toms L-ML, Harden FA, Symons RK, et al. (2007) Polybrominateddiphenyl ethers (PBDEs) in human milk from Australia. *Chemosphere* 68: 797-803.
172. Tue NM, Sudaryanto A, Minh TB,et al. (2010) Accumulation of polychlorinated biphenyls and brominated flame retardants in breast milk from women living in Vietnamese e-waste recycling sites. *Sci Total Environ* 408: 2155-2162.
173. UNEP. Analysis of persistent organic pollutants in developing countries: lessons learned from laboratory projects. UNEP Chemicals 2006. Available from: http://www.chem.unep.ch/pops/laboratory/report%20lessons%20learned.pdf.
174. Environment Canada. Global Atmospheric Passive Sampling (GAPS) Network, 2013. Available from: http://www.ec.gc.ca/rs-mn/default.asp?lang=En&n=22D58893-1.
175. World Health Organization. Biomonitoring of human milk for persistent organic pollutants (POPs). Available from: http://www.who.int/foodsafety/chem/pops_biomonitoring/en/index.html
176. Darnerud PO, Aune M, Lignell S, et al. (2006) Levels of POPs in human breast milk samples from Northern Province, South Africa: comparison to Swedish levels. *Organohalogen Compd* 68: 476-479.

13

Traffic-related air pollution and brain development

Nicholas Woodward , Caleb E. Finch and Todd E. Morgan

Davis School of Gerontology and the Dornsife College, Department of Biological Sciences, University of Southern California, 3715 McClintock Ave, Los Angeles CA 9009-0191, USA

Abstract: Automotive traffic-related air pollution (TRP) imposes an increasing health burden with global urbanization. Gestational and early child exposure to urban TRP is associated with higher risk of autism spectrum disorders and schizophrenia, as well as low birth weight. While cardio-respiratory effects from exposure are well documented, cognitive effects are only recently becoming widely recognized. This review discusses effects of TRP on brain and cognition in human and animal studies. The mechanisms underlying these epidemiological associations are studied with rodent models of pre- and neonatal exposure to TRP, which show persisting inflammatory changes and altered adult behaviors and cognition. Some behavioral and inflammatory changes show male bias. Rodent models may identify dietary and other interventions for neuroprotection to TRP.

Keywords: brain; body weight; prenatal; traffic-related air pollution; inflammation; PM2.5; ultrafine PM

---

1. Introduction

Air pollution from fossil fuel combustion is increasingly recognized for its globally adverse effects on health throughout life. We focus here on traffic-related air pollution (TRP) from roadways in urban settings, for which there is strong epidemiological association with cardiovascular and pulmonary morbidity and mortality [1-4]. Moreover, TRP shows increasing evidence for impact on the brain. Thus, pre- and neonatal developmental exposure to TRP increases risk for low birth weight, and numerous cognitive detriments, including autism spectrum disorders (ASD), schizophrenia, delayed development and cognitive impairment [5-8]. The risks for these disorders are especially high during development, and pollution exposure during gestation can alter development and create lifelong deficits.

First, we discuss epidemiological data on the effects of TRP exposure during gestational development, including impaired fetal growth, as this is often associated with cognitive defects, and the effects of TRP exposure on later life cognition are reviewed. Rodent models are evaluated, as well as the relevant data produced from those models in this nascent field.

2. Overview of TRP

We review two broad groups of TRP: the airborne particulate matter (PM) and the vapor (gaseous) phase, with emphasis on the particulate matter component of TRP. Urban TRP PM is a complex and heterogeneous mixture that includes residues from fossil fuel combustion, organic chemicals, trace metals, nitrate, and sulfate. There are also airborne components from brake linings and the vehicular chassis, as well as roadway components and dust. The recognized size classes of airborne PM range from coarse PM (> 10 μm diameter) to microscopic classes with aerodynamic diameters less than 2.5 μm ($PM_{2.5}$) and 0.1 μm, ($PM_{0.1}$). For each class, primary emissions are transformed from exposure to sunlight and atmospheric ozone and nitric oxides during diurnal and seasonal cycles. While coarse PM are largely trapped by the upper airways, smaller PM can impact the brain directly from olfactory neurons in the nasal mucosa, as well as by systemic effects from the lower airways [9]. The smaller PM sizes are associated with many pathological effects of air pollution [10,11]. Although some studies lump the two smaller size classes under PM_{10}, all three categories have notable adverse effects, as well as different distributions in space and dispersion characteristics.

Of the three categories of particulate matter, $PM_{2.5}$ (fine PM) has received the most attention, with current US EPA standards [12] of 12 μg/m^3. The EPA has not yet addressed $PM_{0.1}$ (UFP, or ultrafine particulates). This class of TRP warrants attention in public health because of experimental evidence for its greater cytotoxicity [13,14], potentially due to the greater penetration through cell membranes [15]. One reason UFP has not been fully appreciated is due to monitoring technology of PM based on weight and not particle number, where it is a large percentage of the total PM. We note alternate terminology of nanoscale PM, $PM_{0.2}$, which encompasses a larger portion of the ultrafine particles, and is considered alongside UFP in this review. UFP is associated with numerous adverse health effects, and comprises the majority of all PM in combustion-derived exhaust [15]. Their small size facilitates the crossing of physiological barriers, including the blood-brain barrier and the placenta as discussed below [6,9,16].

For discussion of developmental exposure, we briefly note that adverse health effects of air pollution exposure increase with closeness to major road [10,17]. UFP was reduced by approximately 80% at a distance of 150 meters from the roadway [11]. Neither $PM_{2.5}$ nor PM_{10} decreased substantially within 150 meters [18], and are decreased by less than 20% at a distance of 400 m vs. 50 m from a roadway [19]. The rate of dilution of UFP was correlated with increased cardiopulmonary mortality, inversely with distance from roadways [10].

3. Epidemiology

3.1. Low birth weight

Because developmental cognitive defects are often associated with fetal growth retardation, it is important that TRP can impact fetal growth. Associations between air pollution exposure in gestation and impaired fetal growth continue to emerge. In particular, $PM_{2.5}$ is associated with low birth weight

(LBW) (< 2,500 g), preterm birth, and small gestational size [20]. However, the critical period for exposure during pregnancy and threshold for these effects remain undefined. We discuss select large-scale studies of PM exposure during gestation. Also see the comprehensive review of Shah and Balkhair [20].

A large Los Angeles based study (n = 220,528) showed 5% greater risk of LBW from $PM_{2.5}$ exposure, with a range of 2.4 μg/m^3 [7]. Other studies used ultrasound to determine the gestational timing of LBW association with air pollution. The largest of these studies (17,660 pregnancies) showed the most consistent association PM_{10} exposure during days 91–120 of pregnancy, where high PM_{10} correlated with smaller abdominal circumference, heard circumference, and femur length [21]. Though this study did not find association with nitric oxide (NO_2) exposure, other studies associated exposures of NO_2 > 38 μg/m^3 with reduced fetal size, femur length, and biparietal distance, even when high NO_2 was recorded only for weeks 12–20 [22,23]. Other studies associated elevated PM_{10} exposure with preterm birth [23,24,25]. A potential mechanism underlying LBW is oxidative stress from maternal exposure during pregnancy to TRP, e.g. increased placental DNA adducts [26].

Obesity is also showing association with air pollution components, which may contribute to diabetes and the metabolic syndrome [27]. Adults (n = 5,228) exposed to NO_2 showed 17% higher risk of diabetes mellitus in the top vs. lowest quintile, differing by 4 ppb [28]. There are also correlations between PM_{10} exposure and the white blood cell count, a marker for systemic inflammation [29].

3.2. Cognitive changes

Epidemiological studies of TRP show negative associations with adult cognition [5,30,31] and brain development [5,30,31,32]. In particular, pre- and postnatal exposure to urban TRP is correlated with autism spectrum disorders (ASD), schizophrenia, and impaired cognitive development. We briefly summarize these findings.

ASD was associated with local gradients in components of TRP, mainly $PM_{2.5}$. Two studies utilizing the California based CHARGE (Childhood Autism Risks from Genetics and the Environment) database found about 2-fold higher odds ratio (OR), for development of ASD, when living near a freeway during the 3rd trimester, and at delivery (< 309 m defined as near, with > 1,419 m as reference group) [31,33]. Exposure during the first postnatal year was associated with 3-fold higher OR for ASD [31]. $PM_{2.5}$ had an OR of 2.08 for gestational exposure, and 2.12 for exposure in the first year of life [31,33]. Similarly, the Nurses' Health Study, a national sample, showed an OR of 2.0 for prenatal diesel particulate exposure, top vs. bottom quintile ($PM_{2.5}$ 4.40 vs. 0.60 μg/m^3) [34]. Contrarily, a study of Swedish twins did not find association of TRP with ASD (PM_{10} 3.3–4.2 μg/m^3); however, this study measured a broader size range of particles [2]. An analysis of 35 pollution components showed higher OR for ASD after exposure to methylene chloride, quinolone, and styrene, but not after diesel PM, or polycyclic aromatic hydrocarbons (PAHs) [35]. The authors noted that the control group had impairments of speech and language, which may have biased the results towards null findings.

Schizophrenia risk is also sensitive to TRP in top vs. bottom quintiles of urbanicity (population density) during gestation, but not during childhood [30]. A study of traffic volume and urbanicity (household crowding, social stressors) concluded that only traffic volume exposure at birth predicted schizophrenia (OR of 4.40 for the top vs. lowest quintile of traffic exposure) [5]. Both studies agree that only exposure during the gestational period correlated with increased risk.

TRP exposure during development is also associated with subclinical cognitive effects, including lower mental development, increased anxiety and depressive behavior, and attentional problems [32,36,37]. A Spanish national study showed decreased mental development for infants of mothers exposed during pregnancy to elevated NO_2 and benzene [36]. Importantly for potential interventions, this association was attenuated in mothers who self-reported a high intake of antioxidant rich foods. We also note the benchmark study of Perera et al. 2003 [38] on PAH levels for Hispanics and African Americans in New York City, which was the first to utilize personal monitors for PAH levels, with greater precision than citywide measurements. Developmental measurements at birth associated high PAH (> average 3.7 ng/m^3 in maternal blood) with a 9% decrease in birth weight, and a 2% decrease in head circumference. The OR for cognitive developmental delay, at 36 months from PAH exposure during gestation was 2.89, for the top vs. bottom quintile [37]. By age 6–7 years, individuals in the top exposure quintile were more anxious and depressed (OR 1.45), with more attentional problems (OR 1.28, top vs. bottom quintile) [32]. For DSM-IV oriented anxiety problems, the OR was a striking 4.59 [32].

4. Rodent models

4.1. Experimental approaches

Several labs have developed rodent models studying the developmental effects of TRP exposure, but no single paradigm has become widely accepted. The main findings (Tables 1–6) include effects on brain morphology, behavior, inflammatory markers, and neurotransmitters. Four experimental paradigms are currently used (Table 1): direct diesel exhaust inhalation, diesel exhaust particle (DEP) oropharyngeal aspiration, the Concentrated Air Particles delivery System (CAPS), and filter-trapped nano-sized ambient reaerosolized particles. Most studies used inhalation, while one lab used direct oropharyngeal aspiration of DEP.

Diesel exhaust: Pregnant mice were exposed to the whole exhaust stream from a diesel engine, diluted to concentrations ranging from 0.171–3.0 mg/m^3. Auten et al. 2012 [47] and Bolton et al. 2012 [48] utilized a 6.4 hp direct injection single cylinder 320-cc Yanmar L70V diesel generator, operating at a constant 3600 rpm. Yokota et al. 2009 [39] used a 2369-cc diesel engine, operating at 1050 rpm. The exhaust includes volatile gaseous components, notably CO, SO_2, and NO_2. Unlike other exposure paradigms, these particles are not filtered by size, and retain native charges. The direct diesel exhaust paradigm is missing other real world pollutants from vehicular traffic, e.g. rubber from tire erosion, brake lining debris, and reaerosolized dust from roadways. Moreover diesel engine exhaust can represent only one type of vehicle, and the particles are being directly emitted and thus did not undergo the secondary reactions from heat and sun exposure, which develop as a function of time after emission.

Table 1. Air pollution exposure sources.

Exposure study	Methods	Particulate matter size and composition	Volatiles	Non-tailpipe pollution	Secondary aerosols
Diesel exhaust					
Direct from diesel engine Bolton et al. 2012 [48]	Auten et al. 2012 [47]	18–200 μm; extractable organic matter, 39.8%	Yes	No	No
Sugamata et al. 2006 [77], Fujimoto et al. 2005 [46], Yokota et al. 2009 [39], Yokota et al. 2013 [57], Suzuki et al. 2010 [62]	Umezawa et al. 2011 [78]	All size particles	CO, 2.67 ppm; NO_2, 0.23 ppm, SO_2, < 0.01 ppm	No	No
Diesel exhaust particles Bolton et al. 2013 [67], 2014* [51]	Auten et al. 2012 [47] Hougaard et al. 2008** [40]	18–200 μm	No	No	No
Traffic-related air pollution					
Ambient Air Wei et al., in prep. [49]		All size particles	Yes	Yes	Yes
Concentrated Ambient Particle System (CAPS)	Allen et al. 2013 [79], 2014a [53], 2014b [58]	< 100 nm dia.	Yes	Yes	Yes
Filter-trapped nano-sized PM (nPM) Davis et al. 2013 [54]	Morgan et al. 2011 [43]	< 200 nm dia. Filter-bound, re-aerosolized.	No	Yes	Yes

Non-tailpipe pollution includes brake dust, tire erosion, and roadway dust. Secondary aerosols arise from alterations of particulate matter by temperature, sunlight, humidity, ozone etc. * Particles delivered by oropharyngeal introduction. ** Diesel exhaust particles reaerosolized for inhalation.

Diesel exhaust particles (DEP):

Oropharyngeal Aspiration: (Auten et al. 2012 [47]) This is the only non-inhalation paradigm used in the prenatal studies. Diesel exhaust particles are collected from a single cylinder diesel engine, and then 50 μg of diesel exhaust particles (DEP) are suspended in 50 μL of PBS with 0.05% Tween-20 and delivered by oropharyngeal aspiration. Importantly, this delivery method bypasses nasal inhalation.

Reaersolized Inhalation: (Hougaard et al. 2008 [40]) Obtained from the National Institute of Standards and Technology, Standard Reference Material 2975, these particles are obtained from a diesel powered forklift, and are re-aerosolized for inhalation delivery [40,41]. Importantly, the re-aerosolized DEP, like the resuspended DEP for oropharyngeal aspiration, lack gaseous and volatile components. Also, because the particles are suspended in water, they are depleted of insoluble PM. Elimination of insoluble particulate matter is of special relevance, as this includes black carbon and polycyclic aromatic hydrocarbons (PAHs).

CAPS (Concentrated Ambient Particle System): Ultrafine fractioned particulate matter is concentrated next to a roadway for direct real time delivery at 10–20 times ambient concentration [42]. CAPS maintain ambient components, including gases and volatiles, and the native charges of the particles. Importantly, condensing the particles does not alter their natural size distributions, and does not amplify aggregation [13,42]. Limitations of this system are its dependence on the current traffic patterns, which fluctuate diurnally and seasonally.

Filter-trapped nano-sized PM from urban TRP: This paradigm, developed by Constantinos Sioutas at the University of Southern California [43], collects ambient air particulate matter, $PM_{0.2}$, on the roadside next to a high traffic source with a high-volume ultrafine particle sampler on 0.2 μm pore Teflon filters. Collections are made for 4–6 weeks in the fall to encompass the range of temperatures and moisture in Southern California [44,45]. The collection is continuous and includes secondary transformations during the diurnal cycle. Besides combustion products, the sample includes reaerosolized roadway dust, and traces from brake lining and tire erosion that are < 0.2 μm. The filters are sonicated in distilled water to yield a suspension, which is stored frozen until use. The reaerosolized PM has average particle size of 60 nm at a density of 350 $\mu g/m^3$, which is about 25x greater than ambient concentrations of that size range of particle [44,45]. The resuspension lacks ambient gases and is depleted in water insoluble organic species including PAHs and black carbon [43]. We designated these materials as nPM (nanoparticulate matter) in distinction from the size class of ambient UFP in the literature. After re-aerosolization, rodents are exposed to nPM together with ambient pollutants in the exposure room, which are 35–50% below outdoor ambient levels, while control animals have this air filtered by HEPA filters.

In summary, each experimental paradigm represents trade-offs. While DE is the most readily obtained, they are model emissions of one engine type and lack secondary atmospheric transformations of ambient pollution. The CAPS fully capture TRP for PM, gases, and volatiles, but vary diurnally and seasonally. The nPM capture diurnal variations, but the resuspension is deficient in black carbon (BC) and PAH among other water insoluble organic compounds.

4.2. Body weight

Epidemiological findings of air pollution on birth weight are corroborated in some rodent models (Table 2). Mice exposed to 3.0 mg/m^3 DEP had decreased fetal weight as early as gestational day (GD) 18 (equivalent to the human third trimester) [46]. Several studies observed a weight decrease four weeks after birth in mice exposed to DEP (Table 3) [40,47]. Finally, two studies observed a reversal later in life, with increased weight at four months of age in mice [48], and at eight weeks in rats [49]. Decreased weight at weaning [40,47], combined with increased weight later in life [48,49], corresponds with previously documented rebounding in weight after a prenatal stressor [50]. Air pollution also has an additive effect on weight when exposure occurs in utero, followed by a high fat diet (HFD = 45% calories from fat) after birth (Table 4) [48,51]. The prenatal DEP + HFD treated mice showed significant weight gain over high fat diet alone, a compounding effect similar to what was observed for inflammatory responses (increased cytokines, microglial activation) [48,51]. Males in both replicates showed greater weight gain from treatment [48,51]. However, females showed 4.4x greater weight gain for the first experiment [48], and no change in the second [51]. Notably, the first experiment used diesel exhaust inhalation, while the second employed oropharyngeal aspiration of DEP, which excludes gas components. Thus, it is possible that either a species lost in the conversion to suspended diesel particles, or the inhalation delivery route caused the weight gain. The high fat postnatal group showed insulin resistance with elevated serum insulin in the males [48,51]. Females showed only a change in serum leptin, while males did not show any differences from pollution exposure.

Exposure to highly polluted ambient Beijing air (75.3 μg/m^3) caused, worsened lipid profiles and weight gain in both rat mothers and offspring [49]. Pregnant dams had higher low-density lipoprotein (LDL), total cholesterol, triglyceride, and overall weight. Pups had increased weight at eight weeks, and worsened lipid profiles, with increased LDL, total cholesterol, and triglyceride, and decreased in HDL. Rodent models also corroborated mid-life and gestational weight effects [27,48,51,52].

Table 2. Prenatal exposure: body weight, behavior, and cell-molecular responses.

Study	Exposure protocol	Weight, gross changes	Behavior	Cell-molecular changes
Diesel exhaust direct				
Bolton et al 2012 [48], 2013 [66]; GD 18 (♂ + ♀)	0.5 or 2.0 mg/m^3 4 h/d; GD 7–17	Weight at 4 m: ♂—1.1 ♀—NC	Bolton 2012 [48]: Open field activity at 4 m: ♂—0.7 ♀—NC Bolton 2013 [66]: NC in forced swim	CCL2/MCP-1—3.5 CX3CL1/fractaline—1.5
Auten et al 2012 [47] GD 18 (♂ + ♀)	0.5 or 2.0 mg/m^3 4 h/d; GD 9–17	Weight at 4 w: 0.5 mg/m^3 (♂ + ♀) 0.9 2.0 mg/m^3 (♂ + ♀) NC		eotaxin: 4 KC: 6 RANTES: 10 +
Fujimoto et al 2005 [46]; GD 14	0.3, 1.0, or 3.0 mg/m^3 12 h/d; GD 2–13	↑ placental weight (♂, ♀—1.0 mg/m^3) ↓ fetal weight (♂, ♀—3.0 mg/m^3)		
Yokota et al 2013 [57] ♂ only- behavior at 5 w	1.0 mg/m^3 8 h/d; GD 2–17		↓ Retention time on rotating rod Cliff avoidance latency to jump 0.68	
Yokota et al 2009 [39] 5 w-behavior	1.0 mg/m^3 8 h/d; GD 2–17		Spontaneous motor activity: ♂—0.83	
Suzuki et al 2010 [62] 5 w	0.171 mg/m^3, 8 h/d, 5 d/w GD 2–16		↓ Spontaneous locomotor activity	
DEP				
Hougaard et al 2008 [40] —19 w (♂ + ♀)	19 mg/m^3 DEP 1 h/d; GD 9–19	Weight at 4 w: 0.9	No effect in the Morris water maze	No indication of any DNA damage, nor inflammation
CAPS				
Allen et al 2013 [79], Allen et al 2014a [53]	15–240 μg/m^3 4 h/d ; PND 4–7		♂ ↓ Response rates for FI60 (6 mo)	

PND 60	& 10–13 4 h/d; PND 56–60		Novel object performance (6 mo): ♂—0.5, ♀—0.8	
Allen et al 2014b [58] 2 w or 8 w	200,000 particles/cm^3 96 µg/m^3 4 h/d PND 4–7 & 10–13	Lateral ventricle size: PND14: ♂ 3.2 PND55: ♂ 1.8		GFAP PND 14: Hipp: ♂—0.5, ♀—1.9 CC: ♂—NC, ♀—1.5 IBA1 PND 55: Hipp: ♂—NC, ♀—NC AC: ♂—1.3, ♀—NC
Filter-trapped nPM				
Davis et al 2013 [54] —8 m	350 µg/m^3 5 h/d 3 d/w, 10 w		Tail suspension immobility 8 m: ♂—2.6, ♀—NC	PD3—JNK1 (♂, ♀) 0.7 Hipp GLU—NC

Abbreviations: AC, Anterior commissure; CAPS, Concentrated Ambient Particle System; CC, Corpus callosum; CCL2, Chemokine (C-C motif) ligand 2; CX3CL1, Chemokine (C-X3-C motif) ligand 1; DEP, diesel exhaust particles; FI60, Fixed interval reward 60 sec; GFAP, Glial fibrillary acidic protein; GD, Gestation day; GLU, Glutamate; Hipp, Hippocampus; IBA1, ionized calcium binding adaptor molecule 1; JNK-1, c-Jun N-terminal kinase 1; KC, keratinocyte chemoattractant; m, months; MCP-1, monocyte chemotactic protein 1; NC, no change; RANTES, regulated on activation, normal T cell expressed and secreted; w, weeks.

Table 3. Shared responses to traffic related air pollution across multiple experiments.

Shared responses		No Change
Body weight	Auten 2012 [47] ↓ 4 w (DE), Bolton 2012 [48] ↑ 5 m (DE), Fujimoto 2005 [46] ↓ GD 14 (DE), Hougaard 2008 [40] ↓ 4 w (DEP), Sugamata 2006 [77] ↓ 4 w (DE), Umezawa 2011 [78] ↓ 8 w (DE), Wei unpub ↓ [49] 2 w & 3 w (Beijing Air)	Allen 2014a [53], Bolton 2014 [51], Davis 2013 [54], Hougaard 2009 [41], Suzuki 2010 [62], Yokata 2013 [57]
Spontaneous locomotor activity	Bolton 2012 [48] ↓ 5 m (DE), Hougaard 2009 [41] ↓ 8 w (DE), Suzuki 2010 [62] ↓ 5 w (DE), Yokata 2009 [57] ↓ 5 w (DE)	Allen 2013 [79], Allen 2014a [53], Bolton 2014 [51], Davis 2013 [54]
Cortex-dopaminergic	Allen 2014a [53] ↓ 9 m (PND 55 CAPS), Allen 2014b [58] ↑ 2 w & 8 w (PND 14 CAPS), Suzuki 2010 [62] ↑ 5 w (DE), Yokata 2013 [57] ↓ 3 w (DE)	
Microglial activation	Allen 2014a [53] ↑ 9 m (PND14 + 55 CAPS), Allen 2014b [58] ↑ 8 w (CAPS), Bolton 2012 [48] ↑ GD 18 (DE), Bolton 2014 [51] ↑ 6 w (DEP + HFD)	Bolton 2013 [66]

Abbreviations: CAPS, Concentrated Ambient Particle System; DE, diesel exhaust; DEP, diesel exhaust particles; HFD, high fat diet; PND, Postnatal day

4.3. Behavioral changes

Behavioral changes from developmental pollution exposure include cognitive and locomotor deficits (Table 2). Cognitive deficits include depressive symptoms, impaired short-term memory, and decreased response rates for fixed interval sixty-second (FI60) reward tests [53,54]. Mice exposed in utero to nPM showed increased immobility on a tail suspension test, which is a marker for depression [54]. Only males were vulnerable, with a decreased delay until first period of immobility, and 2.6× longer immobility versus control, which implicates activation of the amygdala [55,56]. Correspondingly, prenatal exposure to DEP (maternal oropharyngeal DEP aspiration) increased adult amygdala levels of monoamine neurotransmitters (dopamine, dopamine metabolites, and serotonin) [57]. However, the oropharyngeal DEP model did not alter adult forced swim performance, another marker for depression considered equivalent to tail suspension [48]. This discrepancy could be due to the different age at assessment (8 mo vs. 4 mo), or the different pollutant models.

Impaired short-term memory was observed in neonatal mice exposed to CAPS, from postnatal day (PND) 4–7 and 10–13 and assessed at 2 months of age by the novel object recognition test. In the one-hour posttest, CAPS exposed mice spent more time with the familiar versus novel object, indicative of impaired short-term memory [53]. The FI60 test, a model of impulsivity, showed decreased response and run rates, but only for males [53]. Despite smaller overall response rates, there was no significant difference in learning. In a separate experiment, conducted with the same exposure protocols, a secondary dose of pollution from PND 56–60 caused deficits in a fixed interval waiting for reward test, a classic model of impulsivity control [58]. These experiments included CAPS exposure on postnatal days 4–7 and 10–13. The sensitivity to neonatal exposure is important because neonatal rodent nervous systems are relatively less mature compared to humans [59].

Deficits in short-term memory of neonatal CAPS exposure may be mediated by glutamatergic changes. Glutamate levels in the hippocampus, which is critical for spatial learning and memory, were increased 1.26-fold in the male CAPS exposed mice [58]. Though the effect returned to baseline by eight weeks of age, the transient glutamatergic increase during development could cause persisting effects, including excitotoxicity. Detailed studies of hippocampal circuit functions, e.g. LTP, and synaptic density are needed. Increased inflammatory cytokine levels (Il-6, IL-1b, TNF-a) are also relevant to behavioral deficits through their impact on synaptic plasticity. These cytokines showed complex changes in different brain regions in mice exposed to neonatal CAPS [58,60]. For example, TNF-a modulates glutamate, and potentiates the cell to glutamatergic excitotoxicity [61], which could alter short-term memory later in life.

Locomotor deficits from prenatal exposure to pollutants include decreased spontaneous motor activity and impaired balance (Table 4) [39,48,57,62]. Intriguingly, only males have shown decreased spontaneous motor activity in studies from several labs that include ages from 5 weeks to 5 months [39,60,62]. Decreased spontaneous motor activity at age 2 months in CAPS studies was only observed when paired with a second treatment from PND 56–60 (Table 4) [53]. Balance was impaired on the rotating rod test in prenatally exposed male mice at 5 weeks [57]. These mice also had decreased latency in the cliff avoidance test [57]. The impaired performance on these two balance tests is not attributable to differences in body weight [57]; only Bolton et al. 2012 reported weight differences [51]. We note that a shift from direct exposure to diesel exhaust to the oropharyngeal DEP in the same lab did not to replicate these effects [51].

Table 4. Postnatal exposure, secondary manipulations.

Study (secondary treatment)	Behavior	Weight	Serum	Brain cellular	Brain subcellular
Diesel exhaust-direct					
Bolton et al 2012 [48]—6 m DE + HFD HFD—45% fat, beginning 4 m, for 6 w		Weight ♂—NC ♀—4.4	Leptin: ♂—NC ♀—1.7 Insulin: ♂—6.4 ♀—NC	IBA1: Hypothalamus: ♂—1.7, ♀—1.3 Dentate gyrus: ♂—1.4, ♀—1.5 Amygdala: ♂—1.3, ♀—NC Hipp CA1: ♂—1.3, ♀—NC	
Bolton et al 2013 [66] DEP + Nest Restriction DEP: 50 μg DEP every 3 d, from GD 2–17 NR: E 14–19	Contextual fear recall: ♂—0.4 ♀—NC Anxiety behavior: ♂—1.6 ♀—1.2	No change			GD18 TLR4: ♂—1.3, ♀—NC Casp-1: ♂—1.4, ♀—NC IL-1b: ♂—NC ♀—NC Il-10: ♂—0.8, ♀—1.4
Bolton et al 2014 [51] DEP + High Fat Diet DEP: 50 μg DEP every 3 days, from GD 2–17 HFD: 45% Fat, beginning 4 m, for 9 w	Anxiety: (elevated zero maze) ♂—1.1 ♀—NC	Weight: ♂—1.3 ♀—NC	Insulin: ♂—1.8 ♀—NC	↑ peripheral macrophage infiltration in hypothalamus (♂ 1.2) GFAP Hipp NC	Hipp: CD11b: ♂—1.6, ♀—NC TLR4: ♂—1.2, ♀—NC CXC3CR1: ♂—1.2, ♀—NC
CAPS					
Allen et al 2013 [79]—Allen et al 2014a [53]—9 m CAPS: PND 14 and/or PND 55	Waiting for reward behavior: ♂—↓			Corpus Callosum: IBA1: ♂—3.4, ♀—NC, C/A ♂—3.3, ♀—NC, A/C	Hipp: GLU: ♂—1.6, ♀—0.6, C/C DOPAC: ♂—NC, ♀—1.2, C/C

Abbreviations: A/C, Air/CAPS; C/A, CAPS/Air; CAPS, Concentrated air particle system; Casp-1, Caspase-1; CD11b, Cluster of differentiation 11 beta; CXC3CR1, Chemokine (C-X3-C motif) receptor 1; DOPAC, 3,4-Dihydroxyphenylacetic acid; GFAP, Glial fibrillary acidic protein; GLU, Glutamate; IBA1, Ionized calcium binding adaptor molecule 1; IL-1b, Interleukin-1 beta; IL-10, Interleukin 10; IR, Insulin resistance; HFD, High fat diet; Hipp CA1, Hippocampus, cornu Ammonis area 1; NC, No change; NR, Nest restriction; TLR4, Toll-like receptor 4.

4.4. Gross brain morphology

Gross brain weight has not shown sensitivity to prenatal exposure to nPM [54]. However, one study of neonatal exposure to concentrated ambient TRP reported gross enlargement of the lateral ventricles, particularly in males [63]. The nPM prenatally exposed mice revealed no gross brain abnormalities, but quantitation is needed. Cerebral vasculature has just begun to receive attention in air pollution models. After maternal intranasal exposure to black carbon, young adult mice had focal induction of GFAP in astrocyte endfeet on capillary endothelia and altered arterial macrophage granules [64]. These reports point to structural changes that could underlie cognitive dysfunctions from prenatal exposure.

4.5. Neuronal changes

We note a major gap between the body of epidemiological evidence for TRP associations with brain development and the scant information on neuronal changes in animal models of prenatal TRP exposure. Reports on neuronal changes are scattered among different neurotransmitters, often in different brain regions, giving little cohesion of results (Table 5). Dopamine levels in the cerebral cortex illustrate the diversity. Mice exposed to prenatal diesel exhaust had lower cortical dopamine for males at 3 weeks, but no change at six weeks [57]. However, the same exposure paradigm in a different lab showed increased cortical dopamine at 5 weeks [62]. These studies used > 5-fold different levels of DEP density. The turnover of dopamine, estimated by the ratio of the catabolite DOPAC to dopamine, (DOPAC: DA) was higher in neonatally CAPS exposed male mice at two and eight weeks [58].

Neurotransmitter changes reported for adult rodent TRP exposures have not been borne out by prenatal exposures, potentially indicating different mechanisms. We observed decreased glutamate receptor 1 (GLUR1) in the hippocampus in mice exposed to nPM at age three months [43]. However, prenatal exposure did not alter hippocampus GLUR1 at eight months [54]. Neonatal exposure to DEP transiently increased hippocampal glutamate at two weeks, with return to baseline by eight weeks [58].

Cortical neurons harvested from one day old pups prenatally exposed to nPM showed impaired differentiation and neurite initiation, with fewer stage 3 neurons, compared to controls [54]. These pilot studies give a model for linking developmental exposure to alterations in neurons and glia.

Table 5. Neurotransmitter responses.

Experiment	Yokota 2013 [57]		Suzuki 2010 [62]	Allen 2014a [53]		Allen 2014b [58]			
Exposure Condition	Prenatal exposure		Prenatal exposure	Neonatal Exposure		Neonatal Exposure			
Age	♂ 3 w	♂ 6 w	♂ 5 w	♂ 8 w	♀ 8 w	♂ 2 w	♀ 2 w	♂ 8 w	♀ 8 w
Cortex									
DA	0.6	NC	1.6	NC	NC				
DOPAC	NC	NC	2.0	NC	NC				
DOPAC:DA				NC	NC	NC		NC	

HVA	NC	NC	NC						
3-MT	NC	NC	1.5						
GABA				NC	NC				
GLU				NC	NC				
NE			2.0	NC	NC	NC	NC	1.2	0.9
MHPG			1.4						
NM			1.5						
5-HT	NC	1.3							
5-HIAA	NC	NC							
Hippocampus									
DA			NC						
DOPAC					NC	NC		NC	
DOPAC:DA							1.4		NC
HVA			0.2	NC	NC				
3-MT			NC						
GABA				NC	NC		NC		NC
NE			NC	NC	NC				
MHPG			0.7						
NM			NC						
5-HT	NC	NC					NC		1.3
5-HIAA	NC	NC							
Midbrain									
DA							NC		NC
DOPAC:DA						NC	0.7	NC	NC
HVA			0.7	NC	NC		NC		NC
3-MT			0.7						
NE			NC	NC	NC				
MHPG			0.6						
NM			NC						
5-HIAA				NC	NC				
Striatum									
DA			NC	0.8	NC				
DOPAC			NC	NC	NC				
DOPAC:DA				NC	NC				
HVA			NC	0.9	NC				
3-MT			NC						
NE			NC	NC	NC				
MHPG			0.8						
5-HT	NC	NC		NC	NC				
5-HIAA	NC	NC		0.7	NC				
Cerebellum									
DA			NC						
HVA			NC						
NE			NC						

MHPG	0.8	0.8	0.6			
NM	NC	0.6	NC			
5-HT	NC	NC				
5-HIAA	NC	NC				
Hypothalamus						
DA					NC	NC
DOPAC					NC	NC
DOPAC:DA					NC	NC
NE	NC	1.2			0.7	NC
HVA					NC	NC
5-HT	NC	NC				
5-HIAA	NC	1.4				
Amygdala						
DA	1.5	NC				
DOPAC	1.4	1.3				
HVA	1.4	NC				
3-MT	1.4	NC				
5-HT	1.4	NC				
5-HIAA	1.3	NC				

Fold changes (approximated from figures when not given). Abbreviations: DA, dopamine; D OPAC, 3,4-dihydroxyphenylacetic acid (DA catabolite); GABA, gamma-aminobutyric acid; GLU, glutamate; HVA, homovanillic acid (DA catabolite); MHPG, 4-hydroxy-3-methoxyphenylglycol (NE catabolite); NC, No change; NE, norepinephrine; NM, normetanephrine (NE catabolite); 3-MT, 3-methoxytyramine (NE catabolite); 5-HIAA, 5-hydroxyindoleacetic acid (serotonin catabolite); 5-HT, serotonin.

4.6. Inflammatory changes

Inflammation may be a major mediator of maternal systemic and placental responses to air pollution exposure [60]. Systemic inflammation in the mother increases circulating inflammatory cytokines, influencing the development of the fetus, through methods such as the activation of microglia (Table 6) [48,65]. As noted above for neuronal changes, there is a need for similar protocols across labs in the investigation of inflammatory effects.

Prenatal exposure of mice to diesel exhaust rapidly increased cytokines (IL-1b, IL-6, IL-10, and TNF-a) on GD18 [47,48]. Microglial activation is suggested by increased chemokines CCL2/MCP-1 and CX3CL1/Fractalkine [48]. After CAPS neonatal exposure, proinflammatory cytokines (IL-6, IL-1b, TNF-a) were decreased at 2 weeks in males [58]. However, by eight weeks, a full month after the cessation of exposure, IL-1b and TNF-a rebounded to levels 1.4-fold above control's in midbrain [58]. Females showed a different time course, with little upregulation immediately following exposure, yet still showing delayed increases a month later. This effect was brain region specific: unlike the midbrain, the striatum had lower cytokines at the same times [58]. These sex differences in cytokine responses between different brain regions clearly show that inflammatory effects of pollution exposure must be studied in terms of brain pathways and cannot be generalized to the entire brain.

Glial inflammatory responses are detected by the astrocyte specific GFAP (intermediate filament glial fibrillary acidic protein) and the microglial marker of IBA1 (ionized calcium-binding adapter molecule 1). These responses were observed in neonatal exposures, and prenatal exposures compounded with a secondary stimulus. Neonatal mouse exposures from PND 4–7 and 10–13 increased GFAP in the hippocampus, corpus callosum, and anterior commissure in females, while males responded with decreased GFAP [58]. These measurements were made immediately following exposure, at two weeks. IBA1 was upregulated in the hippocampus and corpus callosum for males, at eight weeks and 9 months, respectively [53,58]. Females showed no change for IBA1. Adult mice of several genotypes showed brain inflammatory responses, with induction of IL-1a and TNFa, and activation of microglia and astrocytes [13].

Inhalation of prenatal diesel exhaust, as well as oropharyngeal aspiration of DEP, did not elicit any changes in GFAP or IBA1 in mice, examined at six months of age [48,51]. However a high fat diet (HFD) starting at 4 months and lasting for six weeks, along with the prenatal DE exposure, increased IBA1, but not GFAP [48]. IBA1 was increased in hypothalamus, dentate gyrus, amygdala, and the CA1 of the hippocampus for males [48]. Females showed changes only in the hypothalamus and dentate gyrus [48]. This upregulation of IBA1 supports the hypothesis of air pollution exposure during gestation as an enhancer for later life environmental insults: the DE+HFD group responded more than either treatment alone, with changes in brain regions that did not change with only one of the treatments (Table 3). The two-hit hypothesis postulates that the first insult primes the system [48,60]. While inflammation may not be activated by a single insult, the second inflammatory challenge may cause a disproportionately larger response.

Table 6. Cytokine changes.

	Allen 2014b [58]				Bolton 2012 [48]
	♂ 2 w	♀ 2 w	♂ 8 w	♀ 8 w	♂ + ♀ GD 18
Cortex					
IL-6				0.5	
IL-1b	NC		NC		
Hippocampus					
IL-6	NC		NC		
IL-1b	NC		NC		
Midbrain					
IL-6		NC		2.1	
IL-1b	0.5	NC	1.4	NC	
TNF-a	0.5	NC	1.5	NC	
Striatum					
IL-6		0.4		NC	
IL-1b	NC		NC		
TNF-a	NC	0.7	NC	NC	
Whole Brain					
IL-6					3.7
IL-1b					5.5
TNF-a					
IL-10					3.5

Abbreviations: GD, Gestational day: NC: No change; PND: Postnatal day.

4.7. Sex differences

Sex differences are apparent in rodent responses to prenatal exposure, with greater male vulnerability observed. For open field activity, only male mice showed deficits [48]. For the tail suspension test, a measure of depressive behavior, only males were responsive, with no effect for females [54]. Further sex differences were shown in secondary treatments (Table 3). Release of inflammatory cytokines is profoundly affected by sex, a trend even more pronounced with the addition of a secondary insult. When prenatal diesel exposure is combined with either six weeks HFD, or nest restriction from gestational day (GD) 14–19, male mice show significant effects in numerous cytokines and chemokines, while females show no changes [51,66]. The combination of prenatal diesel exposure and adult high fat diet increased serum insulin, insulin resistance, and IL-1b, again only in male mice [51]. Brain inflammatory proteins CD11b, TLR4, and CXC3CR1 were increased only in males [51]. Likewise, only males had increased peripheral macrophage infiltration in the hypothalamus [51]. Furthermore, for nest restriction paired with diesel exposure, only males showed decreased contextual fear recall, and changes in brain TLR4, caspase-1, IL-1b, and IL-10 [28]. Finally, in the combination of PND 4–7, 10–13, and 55–60 CAPS exposure, only males showed increased IBA1 staining in the corpus callosum [53]. These experimental findings demonstrate greater male vulnerability to pollution exposure, especially when combined with a second insult.

4.8. Nanoparticles in utero

One hypothesis for the mechanism underlying air pollution's physiological effects is unique to the nanoscale particulate matter component. It is possible that the particles might be able to cross the placental barrier to directly interact with the fetus, which develops a blood-brain barrier by gestational day 16 [67]. There is experimental evidence for transplacental transfer of nano-size PM. Titanium dioxide nanoparticles (25–70 nm dia) subcutaneously delivered to pregnant mice on GD 3, 7, 10 and 14, were detected in male brains and testes six weeks postnatally [68]. Thus, model nanoparticles can cross both the maternal placenta and the blood-brain barrier of the developing fetus. *Ex vivo* models with human placenta and polystyrene beads show strong size dependency, with 50 and 80 nm beads rapidly crossing the placenta, possibly by simple diffusion; larger beads > 240 nm do not cross [69]. However, this *ex vivo* model does not represent potential modification of PM by proteins and lipids, which create a bio-corona, altering the movement of the nanoparticles [70]. Engineered nanoparticles of this size show 1000-fold range of translocation to the brain (0.00006% to 0.03%) [63,71]. Although particles may cross secondary barriers (placenta, blood-brain), their inhalation or ingestion does not necessarily allow transport to these secondary barriers. This is of relevance because most experiments inject the particles into the animal, bypassing the lungs. The small size of nanoparticles is important, as particles < 34 nm rapidly translocate from lung to mediastinal lymph node [72]. Notable, negatively charged particles accumulated in secondary organs more than positively charged particles [73].

The placenta may be more vulnerable to nanoparticle entry later in gestation, when the placental wall has thinned and is more vascularized, but also early in gestation before the placenta is fully formed [50]. The period after the placenta is formed, but before maternal-fetal circulatory systems are fully developed, could be less vulnerable to pollution exposure, due to minimal blood flow to the fetus. Nanoparticles may even cause fetal damage without penetrating the placenta, e.g. *in vitro,* nanoparticles can cause DNA damage even when they do not cross a cell barrier [74]. We note that these are not exclusionary hypotheses, and both may potentially be occurring.

4.9. Protective measures

In the urban realities of 21st Century populations, it is not possible to prevent prenatal exposure to TRP by restricting household or school proximity to roadways. Thus, we must consider other means to limit detrimental effects of prenatal TRP exposure. Diet optimization may be a pragmatic approach. As a precedent, higher maternal consumption of fruits and vegetables was associated with better mental development in exposure to benzene and NO_2 [36]. Supplementation with anti-oxidants, e.g. the omega-3 fatty acid, docosahexaenoic acid, may blunt TRP induced oxidative stress [75]. While such interventions could attenuate effects, real solutions must come from technical advancements to lower the use of fossil fuels. Combustion engines can be made more efficient, while petroleum itself can be replaced by methanol, which produces fewer particles during combustion [76].

5. Conclusions

Traffic related air pollution is correlated with numerous detrimental health outcomes: increased cardiovascular mortality from adulthood exposures, and low birth weight and cognitive disorders from gestational exposure. These epidemiological observations are largely verified in animal models. The field is emergent with only a handful of labs worldwide studying animal models. There are huge gaps in the understanding of how TRP affects the brain. Specifically, neuronal changes, both in protein and morphology, and potential epigenetic modifications, are lacking. Another challenge comes from the different experimental paradigms between labs, particularly the source of PM and delivery regimen. Relatively few observations have been corroborated across labs. Even so, there is a clear trend for numerous adverse cognitive effects from pollution exposure during development, and future studies hold many promises.

Acknowledgement

Funding: R21 AG-040683, R21 AG-040753, T32AG0037.

Conflict of interest

All authors declare no conflicts of interest in this paper.

References

1. Brook RD, Rajagopalan S, Pope CA, et al. (2010) Particulate matter air pollution and cardiovascular disease an update to the scientific statement from the American Heart Association. *Circulation* 121: 2331-2378.
2. Gong T, Almqvist C, Bölte S, et al. (2014) Exposure to Air Pollution From Traffic and Neurodevelopmental Disorders in Swedish Twins. *Twin Res Hum Genet* 17: 1-10.

3. Pope CA, Burnett RT, Thurston GD, et al. (2004) Cardiovascular mortality and long-term exposure to particulate air pollution: epidemiological evidence of general pathophysiological pathways of disease. *Circulation* 109: 71-77.
4. Stölzel M, Breitner S, Cyrys J, et al. (2007). Daily mortality and particulate matter in different size classes in Erfurt, Germany. *J Expo Sci Environ Epidemiol* 17: 458-467.
5. Pedersen CB, Raaschou-Nielsen O, Hertel O, et al. (2004) Air pollution from traffic and schizophrenia risk. *Schizophr Res* 66: 83-85.
6. Pietroiusti A, Campagnolo L, Fadeel B (2013) Interactions of engineered nanoparticles with organs protected by internal biological barriers. *Small* 9: 1557-1572.
7. Wilhelm M, Ghosh JK, Su J, et al. (2012) Traffic-related air toxics and term low birth weight in Los Angeles County, California. *Environ Health Perspect* 120: 132-138.
8. Williams K, Alvarez X, Lackner AA (2001) Central nervous system perivascular cells are immunoregulatory cells that connect the CNS with the peripheral immune system. *Glia* 36: 156-164.
9. Oberdörster G, Sharp Z, Atudorei V, et al. (2004) Translocation of inhaled ultrafine particles to the brain. *Inhal Toxicol* 16: 437-445.
10. Hoek G, Brunekreef B, Goldbohm S, et al. (2002). Association between mortality and indicators of traffic-related air pollution in the Netherlands: a cohort study. *Lancet* 360: 1203-1209.
11. Zhou Y, Levy JI (2007) Factors influencing the spatial extent of mobile source air pollution impacts: a meta-analysis. *BMC Public Health* 7: 89.
12. Environmental Protection Agency (2013) Federal Register. Rules and Regulations. Vol. 78, No10
13. Campbell A, Oldham M, Becaria A, et al. (2005) Particulate matter in polluted air may increase biomarkers of inflammation in mouse brain. *Neurotox* 26: 133-140.
14. Utell MJ, Frampton MW (2000) Acute Health Effects of Ambient Air Pollution: The Ultrafine Particle Hypothesis. *J of Aerosol Med* 13: 355-359.
15. Zhang Q, Stanier C, Canagaratna M, et al. (2004) Insights into the chemistry of new particle formation and growth events in Pittsburgh based on aerosol mass spectrometry. *Environ Sci Tech* 38: 4797-4809.
16. Wu J, Wang C, Sun J, et al. (2011) Neurotoxicity of silica nanoparticles: brain localization and dopaminergic neurons damage pathways. *ACS Nano* 5: 4476-4489.
17. Peters A, von Klot S, Heier M, et al. (2004) Exposure to Traffic and the Onset of Myocardial Infarction. *N Engl J Med* 351: 1721-1730.
18. Roorda-Knape MC, Janssen NA, De Hartog JJ, et al. (1998) Air pollution from traffic in city districts near major motorways. *Atmos Environ* 32: 1921-1930.
19. Janssen NA, van Vliet PH, Aarts F (2001) Assessment of exposure to traffic related air pollution of children attending schools near motorways. *Atmos Environ* 35: 3875-3884.
20. Shah PS, Balkhair T (2011) Knowledge Synthesis Group on Determinants of Preterm/LBW births. Air pollution and birth outcomes: a systematic review. *Environ Int* 37: 498-516.
21. Hansen CA, Barnett AG, Pritchard G (2008) The effect of ambient air pollution during early pregnancy on fetal ultrasonic measurements during mid-pregnancy. *Environ Health Perspect* 116: 362-369.
22. Iñiguez C, Ballester F, Estarlich M, et al. (2012) Prenatal exposure to traffic-related air pollution and fetal growth in a cohort of pregnant women. *Occup Environ Med* 69: 736-744.

23. van den Hooven EH, Pierik FH, de Kluizenaar Y, et al. (2012) Air pollution exposure during pregnancy, ultrasound measures of fetal growth, and adverse birth outcomes: a prospective cohort study. *Environ Health Perspect* 120: 150-156.
24. Ritz B, Yu F, Chapa G, et al. (2000) Effect of air pollution on preterm birth among children born in southern California between 1989 and 1993. *Epidemiology* 11: 502-511.
25. Ritz B, Yu F, Fruin S, et al. (2002) Ambient air pollution and risk of birth defects in Southern California. *Am J Epidemiol* 155: 17-25.
26. Kannan S, Misra DP, Dvonch JT, et al. (2006) Exposures to airborne particulate matter and adverse perinatal outcomes: a biologically plausible mechanistic framework for exploring potential effect modification by nutrition. *Environ Health Perspect* 114: 1636-1642.
27. Sun Q, Yue P, Deiuliis JA, et al. (2009) Ambient air pollution exaggerates adipose inflammation and insulin resistance in a mouse model of diet-induced obesity. *Circulation* 119: 538-546.
28. Brook RD, Jerrett M, Brook, et al. (2008) The relationship between diabetes mellitus and traffic-related air pollution. *Int J Occup Environ Med* 50: 32-38.
29. Chen JC, Schwartz J (2008) Metabolic syndrome and inflammatory responses to long-term particulate air pollutants. *Environ Health Perspect* 116: 612-617.
30. Marcelis M, Takei N, van Os J (1999) Urbanization and risk for schizophrenia: does the effect operate before or around the time of illness onset? *Psychol Med* 29: 1197-203.
31. Volk HE, Lurmann F, Penfold B, et al. (2013) Traffic-related air pollution, particulate matter, and autism. *JAMA Psych* 70: 71-77.
32. Perera FP, Tang D, Wang S, et al. (2012) Prenatal polycyclic aromatic hydrocarbon (PAH) exposure and child behavior at age 6-7 years. *Environ Health Perspect* 120: 921-926.
33. Volk HE, Hertz-Picciotto I, Delwiche L, et al. (2011) Residential proximity to freeways and autism in the CHARGE study. *Environ Health Perspect* 119: 873-877.
34. Roberts AL, Lyall K, Hart JE, et al. (2013) Perinatal air pollutant exposures and autism spectrum disorder in the children of nurses' health study II participants. *Environ Health Perspect* 121: 978-984.
35. Kalkbrenner AE, Daniels JL, Chen JC, et al. (2010) Perinatal exposure to hazardous air pollutants and autism spectrum disorders at age 8. *Epidemiology* 21: 631-641.
36. Guxens M, Aguilera I, Ballester F, et al. (2012) Prenatal exposure to residential air pollution and infant mental development: Modulation by antioxidants and detoxification factors. *Environ Health Perspect* 1201: 144-149
37. Perera FP, Rauh V, Whyatt RM, et al. (2006) Effect of prenatal exposure to airborne polycyclic aromatic hydrocarbons on neurodevelopment in the first 3 years of life among inner-city children. *Environ Health Perspect* 114: 1287-1292.
38. Perera FP, Rauh V, Tsai WY, et al. (2003)Effects of transplacental exposure to environmental pollutants on birth outcomes in a multiethnic population. *Environ Health Perspect* 111: 201-205.
39. Yokota S, Mizuo K, Moriya N, et al. (2009) Effect of prenatal exposure to diesel exhaust on dopaminergic system in mice. *Neurosci Lett* 449: 38-41.
40. Hougaard KS, Jensen KA, Nordly P, et al. (2008) Effects of prenatal exposure to diesel exhaust particles on postnatal development, behavior, genotoxicity and inflammation in mice. *Part Fibre Toxicol* 5: 3.
41. Hougaard KS, Saber AT, Jensen KA, et al. (2009) Diesel exhaust particles: effects on neurofunction in female mice. *Basic Clin Pharmacol Toxicol* 105: 139-143.

42. Demokritou P, Gupta T, Ferguson S, et al. (2003) Development of a high-volume concentrated ambient particle system (CAPS) for human and animal inhalation toxicological studies. *Inhal Toxicol* 15: 111-129
43. Morgan TE, Davis DA, Iwata N, et al. (2011) Glutamatergic neurons in rodent models respond to nanoscale particulate urban air pollutants in vivo and in vitro. *Environ Health Perspect* 119: 1003-1009.
44. Daher N, Hasheminassab S, Shafer MM (2013) Seasonal and spatial variability in chemical composition and mass closure of ambient ultrafine particles in the megacity of Los Angeles. *Environ Sci Process Impacts* 15: 283-295.
45. Saffari A, Daher N, Shafer MM (2014) Seasonal and spatial variation in dithiothreitol (DTT) activity of quasi-ultrafine particles in the Los Angeles Basin and its association with chemical species. *J Environ Sci Health A Tox Hazard Subst Environ Eng* 49: 441-451.
46. Fujimoto A, Tsukue N, Watanabe M, et al. (2005) Diesel exhaust affects immunological action in the placentas of mice. *Environ Toxicol* 20: 431-440.
47. Auten RL, Gilmour MI, Krantz QT, et al. (2012) Maternal diesel inhalation increases airway hyperreactivity in ozone-exposed offspring. *Am J Respir Cell Mol Biol* 46: 454-460.
48. Bolton JL, Smith SH, Huff NC, et al. (2012) Prenatal air pollution exposure induces neuroinflammation and predisposes offspring to weight gain in adulthood in a sex-specific manner. *FASEB J* 26: 4743-4754.
49. Wei Y, Zhang J, Li Z. Chronic exposure to Beijing's air pollution increases the risk of obesity and metabolic syndrome [unpublished].
50. Ornoy A (2011) Prenatal origin of obesity and their complications: gestational diabetes, maternal overweight and the paradoxical effects of fetal growth restriction and macrosomia. *Reprod Toxicol* 32: 205
51. Bolton JL, Auten RL, Bilbo SD (2014) Prenatal air pollution exposure induces sexually dimorphic fetal programming of metabolic and neuroinflammatory outcomes in adult offspring. *Brain Behav Immun* 37: 30-44.
52. Sun Q, Wang A, Jin X, et al. (2005) Long-term Air Pollution Exposure and Acceleration of Atherosclerosis and Vascular Inflammation in an Animal Model. *JAMA* 294: 3003-3010.
53. Allen JL, Liu X, Weston D, et al. (2014) Developmental exposure to concentrated ambient ultrafine particulate matter air pollution in mice results in persistent and sex-dependent behavioral neurotoxicity and glial activation. *Toxicol Sci* 140: 160-178.
54. Davis DA, Bortolato M, Godar SC, et al. (2013) Prenatal Exposure to Urban Air Nanoparticles in Mice Causes Altered Neuronal Differentiation and Depression-Like Responses. *PLoS ONE* 8: e64128.
55. Dunlop BW, Nemeroff CB (2007) The role of dopamine in the pathophysiology of depression. *Arch Gen Psychiatry* 64: 327-337.
56. Yang TT, Simmons AN, Matthews SC, et al. (2010) Adolescents with major depression demonstrate increased amygdala activation. *J Am Acad Child Adolesc Psychiatry* 49: 42-51.
57. Yokota S, Moriya N, Iwata M, et al. (2013) Exposure to diesel exhaust during fetal period affects behavior and neurotransmitters in male offspring mice. *J Toxicol Sci* 38: 13-23
58. Allen JL, Liu X, Pelkowski S, et al. (2014) Early postnatal exposure to ultrafine particulate matter air pollution: persistent ventriculomegaly, neurochemical disruption, and glial activation preferentially in male mice. *Environ Health Perspect* 122: 939-945.

59. Rice D, Barone Jr S (2000) Critical periods of vulnerability for the developing nervous system: evidence from humans and animal models. *Environ Health Perspect* 108: 511.
60. Block ML, Calderón-Garciduеñas L. (2009) Air pollution: mechanisms of neuroinflammation and CNS disease. *Trends Neurosci* 32: 506-516.
61. Zou JY, Crews FT (2005) TNFα potentiates glutamate neurotoxicity by inhibiting glutamate uptake in organotypic brain slice cultures: neuroprotection by NFκB inhibition. *Brain Res* 1034: 11-24.
62. Suzuki T, Oshio S, Iwata M, et al. (2010) In utero exposure to a low concentration of diesel exhaust affects spontaneous locomotor activity and monoaminergic system in male mice. *Part Fibre Toxicol* 7: 7.
63. Park EJ, Bae E, Yi J, et al. (2010) Repeated-dose toxicity and inflammatory responses in mice by oral administration of silver nanoparticles. *Environ Toxicol Pharmacol* 30: 162-168.
64. Onoda A, Umezawa M, Takeda K, et al. (2014) Effects of maternal exposure to ultrafine carbon black on brain perivascular macrophages and surrounding astrocytes in offspring mice. *PLoS ONE* 9: e94336.
65. Aaltonen R, Heikkinen T, Hakala K, et al. (2005) Transfer of proinflammatory cytokines across term placenta. *Obstet Gynecol* 106: 802-807.
66. Bolton JL, Huff NC, Smith SH, et al. (2013) Maternal stress and effects of prenatal air pollution on offspring mental health outcomes in mice. *Environ Health Perspect* 121: 1075-1082.
67. Stolp HB, Liddelow SA, Sá-Pereira I (2013) Immune responses at brain barriers and implications for brain development and neurological function in later life. *Front Integr Neurosci* 7: 61
68. Takeda K, Suzuki K, Ishihara A, et al. (2009) Nanoparticles transferred from pregnant mice to their offspring can damage the genital and cranial nerve systems. *J Health Sci* 55: 95-102
69. Wick P, Malek A, Manser P, et al. (2010) Barrier capacity of human placenta for nanosized materials. *Environ Health Perspect* 118: 432-436.
70. Nel AE, Mädler L, Velegol D, et al. (2009) Understanding biophysicochemical interactions at the nano-bio interface. *Nat Mater* 8: 543-557.
71. Kim YS, Song MY, Park JD, et al. (2010) Subchronic oral toxicity of silver nanoparticles. *Part Fibre Toxicol* 7: 20.
72. Choi HS, Ashitate Y, Lee JH, et al. (2010) Rapid translocation of nanoparticles from the lung airspaces to the body. *Nat Biotechnol* 28: 1300-1303.
73. Schleh C, Semmler-Behnke M, Lipka J, et al. (2012) Size and surface charge of gold nanoparticles determine absorption across intestinal barriers and accumulation in secondary target organs after oral administration. *Nanotoxicology* 6: 36-46.
74. Sood A, Salih S, Roh D, et al. (2011) Signalling of DNA damage and cytokines across cell barriers exposed to nanoparticles depends on barrier thickness. *Nature Nanotechnol* 6: 824-833.
75. Bilbo SD, Schwarz JM (2012) The immune system and developmental programming of brain and behavior. *Front Neuroendocrinol* 33: 267-286.
76. Yusof AF, Mamat R, Hafzil M (2013) Comparative study of particulate matter (PM) emissions in diesel engine using diesel-methanol blends, *Appl Mech Mater* 465: 1255-1261.
77. Sugamata M, Ihara T, Sugamata M, et al. (2006) Maternal exposure to diesel exhaust leads to pathological similarity to autism in newborns. *J Health Sci* 4: 486-488

78. Umezawa M, Sakata C, Tanaka N, et al. (2011) Pathological study for the effects of in utero and postnatal exposure to diesel exhaust on a rat endometriosis model. *J Toxicol Sci* 36: 493-498.
79. Allen JL, Conrad K, Oberdörster G, et al. (2013) Developmental exposure to concentrated ambient particles and preference for immediate reward in mice. *Environ Health Perspect* 121: 32-38.

14

Monitoring serum PCB levels in the adult population of the Canary Islands (Spain)

Guillermo Burillo-Putze [1,†], Luis D. Boada [2,†], Luis Alberto Henríquez-Hernández [2], Patricio Navarro [3], Manuel Zumbado [2], María Almeida-González [2], Ana Isabel Dominguez-Bencomo [1], María Camacho [2] and Octavio P. Luzardo [2]

[1] Grupo de Investigación en Toxicología Clínica, Hospital Universitario de Canarias, Santa Cruz de Tenerife, España, Facultad de Ciencias de la Salud, Universidad Europea de Canarias, La Orotava, Tenerife, España

[2] Toxicology Unit, Clinical Sciences Department, Universidad de Las Palmas de Gran Canaria, Centro de Investigación Biomédica en Red de Fisiopatología de la Obesidad y Nutrición (CIBER ObN) and Instituto Canario de Investigación del Cáncer (ICIC) Plaza Dr. Pasteur s/n 35016- Las Palmas de Gran Canaria, Spain

[3] Urology Service, Complejo Hospitalario Universitario Insular-Materno Infantil, Avda. Marítima del Sur, 35016 - Las Palmas de Gran Canaria, and Instituto Canario de Investigación del Cáncer (ICIC), Spain

[†] The first and second authors have contributed equally to this work and, therefore, they should be considered indistinctly as first authors.

Abstract: Polychlorinated biphenyls (PCBs) are persistent organic chemicals that have been detected in human serum or tissues all over the world. These pollutants could exert a number of deleterious effects on humans and wildlife, including carcinogenic processes. The Spanish population of the Canary Islands was evaluated with respect to PCB levels more than ten years ago showing lower levels than other Western populations. The objective of our study was to assess the current level of contamination by PCBs showed by this population. We measured serum PCBs in a sample of healthy adult subjects (206 serum samples from subjects with an average age of 66 years old) to evaluate the potential modification of PCB serum levels in this population during the last decade. PCB congeners (28, 52, 77, 81, 101, 105, 114, 118, 123, 126, 138, 153, 156, 157, 167, 169, 180, and 189) were measured by gas chromatography-mass spectrometry (GC/MS). Our results showed that PCB residues were found in 84% of serum samples analyzed, the congeners 28, 153 and 180 being the most frequently detected and at the highest median values (0.1 ng/mL). In addition, the median concentration of the sum of those PCBs considered as markers of environmental contamination by these chemicals (Marker-PCBs) was 0.6 ng/mL, reaching values as high as as 2.6 ng/mL in the 95th percentile. Levels of the sum of PCBs with toxic effects similar to dioxins (dioxin-like PCBs) reached median values of 0.4 ng/mL in the 95th percentile. The reported levels are similar to those described previously in this population more than ten years ago, in the sense that the inhabitants of the Canary Archipelago show levels of PCB contamination lower than the majority of populations from developed countries. These findings suggest that currently there is not any active source of these chemicals in this archipelago. Nevertheless, as foods seem to be a relevant source for these compounds, Public Health authorities should monitor the presence of PCB residues in foods available in the market of these Islands.

Keywords: polychlorobiphenyls; POPs; serum samples; adult subjects; biomonitoring; gas chromatography/mass spectrometry

1. Introduction

Polychlorinated biphenyls (PCBs) are halogenated aromatic hydrocarbons widely used in closed systems such as electrical transformers, and capacitors, as well as in a large number of other applications: paints, fire retardants, etc [1,2]. The volatility of these compounds results in their evaporation from water surfaces and their movement through the atmosphere, resulting in widespread dispersal in the environment [2]. Furthermore, due to their fat solubility and resistance to chemical and biological degradation, ingestion of PCBs by animals leads to bioaccumulation throughout their lives and also to biomagnification in the food chain [1]. Due to these characteristics PCBs are considered as persistent organic pollutants (POPs) and were included in the Stockholm Convention [3].

PCB production decreased and eventually ceased in the 1970s [4]. In fact, they were banned in most Western countries (among them, Spain) in the late 1980s [5]. Currently, in addition to their ubiquitous presence in the environment, PCBs could also reach soils and waters as a result of leaks, spills, and improper disposal of old PCB-containing equipment. Furthermore, foods, mainly those of animal origin, are considered currently as a relevant source of PCBs for Western populations [6].

PCB exposure has been associated with adverse effects on human health. Thus, it has been found that PCB exposure could be related with the incidence of neurocognitive and endocrine disorders [7-10], and PCBs are considered as probable carcinogens to humans (Group 2A) by the International Agency for Research on Cancer (IARC), especially those PCB congeners that exhibit toxic actions similar to dioxins named dioxin-like PCBs (DL-PCBs) [11,12]. Additionally, several PCB congeners are known to be estrogenic and have been considered as endocrine disrupters [13].

The population of the Canary Islands in the context of a Nutritional Survey was extensively studied in 1998 about its level of contamination by POPs, including organochlorine pesticides [14-16]. Those studies indicated that the inhabitants of the Canary Archipelago showed levels of contamination by PCBs lower than the majority of populations from developed countries. In order to asses if there was any variation in POPs levels over the past decade, we have measured PCB serum levels in healthy adults enrolled as controls in a case-control study for bladder cancer developed in these Islands.

2. Materials and Method

2.1. Study area

The Canary Islands are located 1,600 kilometers away from southwest Spain, in the Atlantic Ocean, and hardly 100 kilometers from the nearest point of the North African coast (southwest of Morocco). Geographically, the Islands are part of the African continent; yet from a historical, economic, political and socio-cultural point of view, the Canaries are completely European. The Archipelago consists of seven major islands (the two capital Islands, Gran Canaria and Tenerife; and five other Islands, Fuerteventura, Lanzarote, La Palma, La Gomera, and El Hierro). The economy of the Canary Islands is based fundamentally on a few economic sectors: tourism and to a much lesser extent farming, livestock production, and fishing. Other economic sectors such as traditional polluting industries have a very limited presence in the Islands [14].

2.2. Study group and sample collection

The study population was the healthy controls (206 adult subjects, 38 women and 168 men residents in these Islands at least during the last ten years) enrolled in a case-control study developed in the most populated island of the Canary Archipelago (Gran Canaria Island). All the details about the selection and criteria of inclusion in the case-control study have been described elsewhere [16].

The subjects included in the control group were contacted and asked to participate in the study, to sign the informed consent document, and to complete the specific questionnaire developed *ad hoc* (including questions about the most widely known POPs-related lifestyle). Thus, the questionnaire collected data on professional activities, and dietary habits. The study was approved by the Research Ethics Committee of the Complejo Hospitalario Universitario Insular Materno-Infantil. The participants had blood samples extracted after 12-hour fasting in order to determine the presence of

Table 1. Characteristics of the study population (*n* = 206).

	N (%)
Gender	
Male	168 (81.6)
Female	38 (18.4)
Habitat	
Rural <10,000 inhabitants	7 (3.4)
Semi-rural 10-100,000 inhabitants	38 (18.4)
Urban > 100,000 inhabitants	161 (78.2)
	Mean ± SD
Age (years old)	66.7±12.1
BMI (kg/m^2)	26.3±3.3

residues of POPs. The PCBs were measured in 206 subjects. The sociodemographic characteristics of the study subjects are shown in Table 1.

2.3. Sample preparation and analytical procedures

The aliquots of serum were subjected to solid-phase extraction as described previously [5] and analyzed by gas chromatography/mass spectrometry (GC/MS) using appropriate internal standards [5]. The analytes included in this study were the PCB congeners with IUPAC numbers #28, 52, 77, 81, 101, 105, 114, 118, 123, 126, 138, 153, 156, 157, 167, 169, 180, and 189. Chromatographic analysis was performed using a Thermo-Finnigan TRACE DSQ GC/MS instrument as previously reported [17,18]. We considered the limit of quantification (LOQ) as 10-fold the standard deviation of the blank. The LOQ for PCB congeners 28, 52, 101, 118 and 138 was 0.010 ng/mL; for PCB congeners 153 and 180, it was 0.005 ng/mL. The LOQ for the rest of analytes was 0.001 ng/mL.

In this work we express the total PCB body burden (∑PCBs) as the sum of the 18 PCBs measured (IUPAC congeners #28, 52, 77, 81, 101, 105, 114, 118, 123, 126, 138, 153, 156, 157, 167, 169, 180 and 189), Marker PCB body burden (∑M-PCBs) as the sum of those congeners considered as markers of environmental contamination for PCBs (IUPAC congeners #28, 52, 101, 118, 138, 153 and 180), non-ortho dioxin-like PCBs as the sum of the 4 non-ortho congeners measured (IUPAC congeners #77, 81, 126 and 169), mono-ortho dioxin-like PCBs as the sum of the eight mono-ortho congeners measured (IUPAC congeners #105, 114, 118, 123, 156, 157, 167 and 189), and DL-PCBs body burden as the sum of the 12 DL-PCBs (IUPAC congeners #77, 81, 105, 114, 118, 123, 126, 156, 157, 167, 169 and 189) measured.

2.4. Statistical analysis

Database management and statistical analysis were performed with PASW Statistics v 22.0 (SPSS Inc., Chicago, IL, USA). As serum PCB levels do not follow a normal distribution, the results are expressed with the median, and the 5th and 95th percentiles (p5 and p95, respectively). In addition, a zero value was assigned to all the values below LOQ, as previously reported [5]. Differences in the PCB levels between two groups or more were tested with the non-parametric Mann-Whitney U-test and Kruskal Wallis test. The categorical variables are presented as percentages and were compared between variables with the chi-squared test. P value of less than 0.05 (two-tail) was considered to be statistically significant.

3. **Results and Discussion**

As shown in Table 2, in our series we note that the frequency of detection of the M-PCBs was high, with more than 80% of the study population showing detectable values of this type of PCBs. No significant differences in PCB values among PCBs (individually or grouped) and the rest of variables recorded (gender, habitat, age) were found.

It has to be highlighted that congeners 28, 138, 153 and 180, were detected in a similar percentage of subjects (more than 70% of the adult subjects included in the study). These results agree with our results described previously in that population [19]. Except in the case of PCB-28, which was detected in a much lower proportion in our previous study (6% of the samples). In addition, it has drawn our attention that PCB-28 was also found at concentrations as high as 0.5 ng/mL in the subjects included in percentile 95.

As shown in Table 2, the most toxic and carcinogenic compounds, that is PCBs similar to dioxins, were detected near to 70% of the subjects. This result is of concern because of the adverse health effects attributed to DL-PCBs [20]. Even more so, having into account that those subjects included in percentile 95 may show levels of these carcinogens as high as 0.4 ng/mL.

Table 2. PCB serum levels (ng/mL) and frequency of detection found in the adult subjects enrolled in the present study (average age 66 years old; *n* = 206). In addition data from the previous study developed in 1998 are shown (average age 65 years old; *n* = 116).

		Present series		Series of 2011
Group of PCBs	**Congener**	**Frequency (%)**	**Median (p5-p95)**	**Median (p5-p95)**
Marker-PCBs		**84.0**	**0.6 (0.0-2.6)**	**0.6 (0.0-3.0)**
	PCB-28	76.7	0.1 (0.0-0.5)	0.0 (0.0-0.01)
	PCB-52	39.8	0.0 (0.0-0.1)	0.0 (0.0-0.0)
	PCB-101	27.2	0.0 (0.0-0.02)	0.0 (0.0-0.06)
	PCB-118	65.0	0.0 (0.0-0.3)	0.0 (0.0-0.2)
	PCB-138	74.3	0.0 (0.0-0.5)	0.0 (0.0-1.3)
	PCB-153	74.8	0.1 (0.0-0.8)	0.3 (0.0-1.3)
	PCB-180	71.4	0.1 (0.0-0.8)	0.1 (0.0-0.3)
DL-PCBs (non-ortho)		**16.0**	**0.0 (0.0-0.002)**	**0.0 (0.0-0.4)**
	PCB-77	1.5	0.0 (0.0-0.0)	0.0 (0.0-0.0)
	PCB-81	4.4	0.0 (0.0-0.01)	0.0 (0.0-0.4)
	PCB-126	9.2	0.0 (0.0-0.02)	0.0 (0.0-0.0)
	PCB-169	1.9	0.0 (0.0-0.0)	0.0 (0.0-0.0)
DL-PCBs (mono-ortho)		**67.5**	**0.0 (0.0-0.4)**	**0.0 (0.0-0.4)**
	PCB-105	1.5	0.0 (0.0-0.0)	0.0 (0.0-0.0)
	PCB-114	6.3	0.0 (0.0-0.02)	0.0 (0.0-0.0)
	PCB-118	65.0	0.0 (0.0-0.3)	0.0 (0.0-0.2)
	PCB-123	6.3	0.0 (0.0-0.02)	0.0 (0.0-0.0)
	PCB-156	8.3	0.0 (0.0-0.03)	0.0 (0.0-0.2)
	PCB-157	15.5	0.0 (0.0-0.05)	0.0 (0.0-0.0)
	PCB-167	0.5	0.0 (0.0-0.0)	0.0 (0.0-0.6)
	PCB-189	6.8	0.0 (0.0-0.01)	0.0 (0.0-0.0)
Total DL-PCBs		**67.5**	**0.0 (0.0-0.4)**	**0.0 (0.0-0.7)**
Total PCBs		**84.0**	**0.6 (0.0-2.6)**	**0.7 (0.0-3.9)**

As in our previous works, the present findings indicate that serum PCB levels from people living in the Canary Islands are lower than those reported for other populations, probably due to the fact that there is a limited presence of industrial sources of PCBs in these Islands [19].

These low levels of contamination by PCBs could be due to the dietary patterns of the population under study because diet seemed to be the major source of these chemicals for populations all over the world. A number of studies have reported extensive data on concentrations of PCBs in food samples [18,19,21,22].

Nowadays, it is considered that the consumption of fish is likely to be the most relevant source of PCBs for human beings [22-24]. Fish consumption is lower in the population of the Canary Islands [25] compared to the mainland Spanish population [26], which could at least partially explain the low PCB body burden in the Canary population. Nevertheless, further studies on this aspect are necessary to evaluate the role played by dietary habits as determinants of serum PCB levels.

4. Conclusions

In the present study, concentrations of PCBs were measured in the serum of adult healthy people from the Canary Islands. Our results show that the overall levels of these persistent pollutants are low as described previously for this population, suggesting that e main sources of PCBs for this population (industrial activity and food), remain little relevance in this Islands. In any case, the contribution of PCBs through food may help to explain why these levels have not decreased clearly in the last ten years.

Acknowledgements

This work was supported by FUNCIS PI 35/08 and of the Gobierno de Canarias (Spain), and PI 19/08 of the Instituto Canario de Investigación del Cáncer.

Conflict of Interest

All authors declare no conflicts of interest in this paper.

References

1. Safe SH (1994) Polychlorinatedbiphenyls (PCBs): environmental impact, biochemical and toxic responses, and implications for risk assessment. *Crit Rev Toxicol* 24: 87-149.
2. Headrick ML, Hollinger K, Lovell RA, et al (1999) PBBs, PCBs, and dioxins in food animals, their public health implications. *Vet Clin North Am Food Anim Pract* 15:109-131.
3. United Nations Environmental Program (2001) "Stockholm Convention on persistent organic pollutants. Available from: http://chm.pops.int/Programmes/NewPOPs/tabid/672/language/en-US/Default.aspx
4. Tanabe S (1988) PCB problems in the future: foresight from current knowledge. *Environ Pollut* 50: 5-28.
5. Henriquez-Hernandez LA, Luzardo OP, Almeida-Gonzalez M, et al. (2011) Background levels of polychlorinated biphenyls in the population of the Canary Islands (Spain). *Environ Res* 111: 10-16.

6. Rodriguez-Hernández A, Boada LD, Almeida-González M, et al. (2015) An estimation of the carcinogenic risk associated with the intake of multiple relevant carcinogens found in meat and charcuterie products. *Sci Total Environ* 514C: 33-41.
7. Hagmar L (2003) Polychlorinated biphenyls and thyroid status in humans: a review. *Thyroid* 13: 1021-1028.
8. Longnecker MP, Wolff MS, Gladen BC, et al. (2003) Comparison of polychlorinated biphenyl levels across studies of human neurodevelopment. *Environ Health Perspect* 111: 65-70.
9. Fitzgerald EF, Belanger EE, Gomez MI, et al. (2008) Polychlorinated biphenyl exposure and neuropsychological status among older residents of upper Hudson River communities. *Environ Health Perspect* 116: 209-215.
10. Philibert A, Schwartz H, Mergler D (2009) An exploratory study of diabetes in a first nation community with respect to serum concentrations of *p,p'*-DDE and PCBs and fish consumption. *Int J Environ Res Public Health* 6: 3179-3189.
11. IARC (2002) "IARC Monographs on the evaluation of carcinogenic risk to humans". Available from: http://monographs.iarc.fr/ENG/Monographs/vol82/mono82.pdf
12. Van denBerg M, DeJongh J, Poiger H, et al. (1994).The toxicokinetics and metabolism of polychlorinated dibenzo-p-dioxins (PCDDs) and dibenzofurans (PCDFs) and their relevance for toxicity. *Crit Rev Toxicol* 24: 1-74.
13. Buck Louis GM, Dmochowski J, Lynch C, et al. (2009) Polychlorinated biphenyl serum concentrations, lifestyle and time-to-pregnancy. *Hum Reprod* 24: 451-458.
14. Zumbado M, Goethals M, Álvarez-León EE, et al. (2005) Inadvertent exposure to organochlorine pesticides DDT and derivatives in people from the Canary Islands (Spain). *Sci Total Environ* 339: 49-62.
15. Luzardo OP, Goethals M, Zumbado M, et al. (2006) Increasing serum levels of non-DDT-derivative organochlorine pesticides in the younger population of the Canary Islands (Spain). *Sci Total Environ* 367: 129-138.
16. Boada LD, Henríquez-Hernández LA, Navarro P, et al. (2015) Exposure to polycyclic aromatic hydrocarbons (PAHs) and bladder cancer: evaluation from a gene-environment perspective in a hospital-based case-control study in the Canary Islands (Spain). *Int J Occup Environ Health* 21: 23-30.
17. Luzardo OP, Mahtani V, Troyano JM, et al. (2009) Determinants of organochlorine levels detectable in the amniotic fluid of Women from Tenerife Island (Canary Islands, Spain). *Environ Res*109: 607-613.
18. Almeida-González M, Luzardo OP, Zumbado M, et al. (2012) Levels of organochlorine contaminants in organic and conventional cheeses and their impact on the health of consumers: an independent study in the Canary Islands (Spain). *Food Chem Toxicol* 50: 4325-4332.
19. Luzardo OP, Almeida-González M, Henríquez-Hernández LA, et al. (2012) Polychlorobiphenyls and organochlorine pesticides in conventional and organic brands of milk: occurrence and dietary intake in the population of the Canary Islands (Spain). *Chemosphere* 88: 307-315.
20. Van denBerg M, Birnbaum LS, Denison M, et al. (2006) The 2005 World Health Organization reevaluation of human and mammalian toxic equivalency factors for dioxins and dioxin-like compounds. *Toxicol Sci* 93: 223-241.
21. Luzardo OP, Rodríguez-Hernández A, Quesada-Tacoronte Y, et al. (2013). Influence of the method of production of eggs on the daily intake of polycyclic aromatic hydrocarbons and

organochlorine contaminants: an independent study in the Canary Islands (Spain). *Food Chem Toxicol* 60: 455-462.

22. Bocio A, Domingo JL, Falco G, et al. (2007) Concentrations of PCDD/PCDFs and PCBs in fish and sea food from the Catalan (Spain) market: estimated human intake. *Environ Int* 33: 170-175.
23. Gomara B, Bordajandi LR, Fernandez MA, et al. (2005) Levels and trends of polychlorinated dibenzo-p-dioxins/furans (PCDD/Fs) and dioxin-like polychlorinated biphenyls (PCBs) in Spanish commercial fish and shellfish products,1995–2003. *J Agric FoodChem* 53: 8406-8413.
24. Bordajandi LR, Martın I, Abad E, et al. (2006) Organochlorine compounds (PCBs, PCDDs and PCDFs) in sea fish and sea food from the Spanish Atlantic Southwest Coast. *Chemosphere* 64: 1450-1457.
25. Serra Majem L (2000) Evaluación del estado nutricional de la población canaria (1997–1998). Sociedad Latinoamericana de Nutrición, Caracas, Venezuela.
26. Welch AA, Lund E, Amiano P, et al. (2002) Variability of fish consumption within the 10 European countries participating in the European investigation into cancer and nutrition (EPIC) study. *Public Health Nutr* 5: 1273-1285.

Permissions

All chapters in this book were first published in AIMSES, by AIMS Press; hereby published with permission under the Creative Commons Attribution License or equivalent. Every chapter published in this book has been scrutinized by our experts. Their significance has been extensively debated. The topics covered herein carry significant findings which will fuel the growth of the discipline. They may even be implemented as practical applications or may be referred to as a beginning point for another development.

The contributors of this book come from diverse backgrounds, making this book a truly international effort. This book will bring forth new frontiers with its revolutionizing research information and detailed analysis of the nascent developments around the world.

We would like to thank all the contributing authors for lending their expertise to make the book truly unique. They have played a crucial role in the development of this book. Without their invaluable contributions this book wouldn't have been possible. They have made vital efforts to compile up to date information on the varied aspects of this subject to make this book a valuable addition to the collection of many professionals and students.

This book was conceptualized with the vision of imparting up-to-date information and advanced data in this field. To ensure the same, a matchless editorial board was set up. Every individual on the board went through rigorous rounds of assessment to prove their worth. After which they invested a large part of their time researching and compiling the most relevant data for our readers.

The editorial board has been involved in producing this book since its inception. They have spent rigorous hours researching and exploring the diverse topics which have resulted in the successful publishing of this book. They have passed on their knowledge of decades through this book. To expedite this challenging task, the publisher supported the team at every step. A small team of assistant editors was also appointed to further simplify the editing procedure and attain best results for the readers.

Apart from the editorial board, the designing team has also invested a significant amount of their time in understanding the subject and creating the most relevant covers. They scrutinized every image to scout for the most suitable representation of the subject and create an appropriate cover for the book.

The publishing team has been an ardent support to the editorial, designing and production team. Their endless efforts to recruit the best for this project, has resulted in the accomplishment of this book. They are a veteran in the field of academics and their pool of knowledge is as vast as their experience in printing. Their expertise and guidance has proved useful at every step. Their uncompromising quality standards have made this book an exceptional effort. Their encouragement from time to time has been an inspiration for everyone.

The publisher and the editorial board hope that this book will prove to be a valuable piece of knowledge for researchers, students, practitioners and scholars across the globe.

List of Contributors

Yifeng Wang
Sandia National Laboratories, P. O. Box 5800, Albuquerque, NM 87185, USA

Carlos F. Jove-Colon
Sandia National Laboratories, P. O. Box 5800, Albuquerque, NM 87185, USA

Robert J. Finch
Sandia National Laboratories, P. O. Box 5800, Albuquerque, NM 87185, USA

Hyunok Choi
Departments of Environmental Health Sciences, Epidemiology, and Biostatistics, University at Albany, School of Public Health, One University Place, Rm 153, Rensselaer, NY 12144-3456, USA

Mark T. McAuley
Chemical Engineering Department, Faculty of Science and Engineering, Thornton Science Park, University of Chester, Chester, CH2 4NU, UK

David A. Lawrence
Wadsworth Center, New york state department of healthCenter for Medical Sciences, Rm 1155, 150 New Scotland Ave. Albany, NY 12201, USA

Volodymyr Ivanov
School of Civil and Environmental Engineering, Nanyang Technological University, 50 Nanyang

Viktor Stabnikov
School of Civil and Environmental Engineering, Nanyang Technological University, 50 Nanyang
Department of Biotechnology and Microbiology, National University of Food Technologies, 68
Volodymyrska Street, Kyiv 01601, Ukraine

Chen Hong Guo
School of Civil and Environmental Engineering, Nanyang Technological University, 50 Nanyang

Olena Stabnikova
School of Civil and Environmental Engineering, Nanyang Technological University, 50 Nanyang

Zubair Ahmed
Department of Civil Engineering, King Abdulaziz University, Jeddah 21589, Kingdom of Saudi Arabia

In S. Kim
School of Environmental Science and Engineering, Gwangju Institute of Science and Technology, 123 Cheomdan-gwagiro, Buk-gu, Gwangju 500712, Republic of Korea

Eng-Ban Shuy
School of Civil and Environmental Engineering, Nanyang Technological University, 50 Nanyang

Jianfeng Wen
American Water, 1025 Laurel Oak Road, Voorhees, New Jersey 08043, USA

Yanjin Liu
American Water, 1025 Laurel Oak Road, Voorhees, New Jersey 08043, USA

Yunjie Tu
American Water, 1025 Laurel Oak Road, Voorhees, New Jersey 08043, USA

Mark W. LeChevallier
American Water, 1025 Laurel Oak Road, Voorhees, New Jersey 08043, USA

Michael R. Templeton
Department of Civil and Environmental Engineering, Imperial College London, London, United Kingdom SW7 2AZ

Acile S. Hammoud
Department of Civil and Environmental Engineering, Imperial College London, London, United Kingdom SW7 2AZ

Adrian P. Butler
Department of Civil and Environmental Engineering, Imperial College London, London, United Kingdom SW7 2AZ

Laura Braun
Department of Civil and Environmental Engineering, Imperial College London, London, United Kingdom SW7 2AZ

Julie-Anne Foucher
Department of Civil and Environmental Engineering, Imperial College London, London, United Kingdom SW7 2AZ

Johanna Grossmann
Department of Civil and Environmental Engineering, Imperial College London, London, United Kingdom SW7 2AZ

Moussa Boukari
Département des Sciences de la Terre, Université d'Abomey-Calavi, Cotonou, Bénin 01 BP 4521

Serigne Faye
Département de Géologie, Université Cheikh Anta Diop, Dakar, Sénégal PO Box 5005

Patrice Jourda
Département des Sciences et Techniques de l'Eau et du Génie de l'Environnement, Université
Félix Houphouët-Boigny, Abidjan, Côte d'Ivoire 22 BP 582

Lars Carlsen
Awareness Center Linkøpingvej 35, Trekroner, DK-4000 Roskilde, Denmark
Center of Physical Chemical Methods of Research and Analysis, al-Farabi Kazakh National
University, 96A Tole Bi st., 050012, Almaty, Kazakhstan

Rafael Gonzalez-Olmos
IQS School of Engineering, Universitat Ramon Llull, Via Augusta, 390, 08017 Barcelona, Spain

Jennifer K. Costanza
North Carolina Cooperative Fish and Wildlife Research Unit, Department of Applied Ecology, North Carolina State University, Raleigh, NC 27695, USA
Department of Forestry and Environmental Resources, North Carolina State University, Raleigh, NC 27695, USA

Robert C. Abt
Department of Forestry and Environmental Resources, North Carolina State University, Raleigh, NC 27695, USA

Alexa J. McKerrow
Core Science Analytics and Synthesis, U.S. Geological Survey, Raleigh, NC 27695, USA

Jaime A. Collazo
U.S. Geological Survey, North Carolina Cooperative Fish and Wildlife Research Unit, Department of Applied Ecology, North Carolina State University, Raleigh, NC 27695, USA

Périne Doyen
Université du Littoral Côte d'Opale, Institut Charles Viollette, Equipe Biochimie des Produits
Aquatiques (BPA), Boulevard du Bassin Napoléon, 62327 Boulogne-sur-Mer, France

Etienne Morhain
Université de Lorraine, UMR CNRS 7360 : Laboratoire Interdisciplinaire des Environnements
Continentaux (LIEC), Rue Delestraint, 57070 Metz, France

François Rodius
Université de Lorraine, UMR CNRS 7360 : Laboratoire Interdisciplinaire des Environnements
Continentaux (LIEC), Rue Delestraint, 57070 Metz, France

Yang Wang
Department of Paper and Bioprocess Engineering, State University of New York—College of Environmental Science and Forestry, Syracuse, NY 13210, USA

Chenhui Liang
Department of Paper and Bioprocess Engineering, State University of New York—College of Environmental Science and Forestry, Syracuse, NY 13210, USA

Shijie Liu
Department of Paper and Bioprocess Engineering, State University of New York—College of
Environmental Science and Forestry, Syracuse, NY 13210, USA

Luigi Falletti
Department of Industrial Engineering, University of Padova, via Marzolo 9-35127 Padova, Italy

Lino Conte
Department of Industrial Engineering, University of Padova, via Marzolo 9-35127 Padova, Italy

Andrea Maestri
Polesine Acque SpA, via B. Tisi da Garofolo 11-45100 Rovigo, Italy

Peter Mochungong
Exposure and Biomonitoring Division, Health Canada, Ottawa, Canada K1A 0K9

Jiping Zhu
Exposure and Biomonitoring Division, Health Canada, Ottawa, Canada K1A 0K9

Nicholas Woodward
Davis School of Gerontology and the Dornsife College, Department of Biological Sciences, University of Southern California, 3715 McClintock Ave, Los Angeles CA 9009-0191, USA

Caleb E. Finch
Davis School of Gerontology and the Dornsife College, Department of Biological Sciences, University of Southern California, 3715 McClintock Ave, Los Angeles CA 9009-0191, USA

Todd E. Morgan
Davis School of Gerontology and the Dornsife College, Department of Biological Sciences, University of Southern California, 3715 McClintock Ave, Los Angeles CA 9009-0191, USA

Guillermo Burillo-Putze
Grupo de Investigación en Toxicología Clínica, Hospital Universitario de Canarias, Santa Cruz de Tenerife, España, Facultad de Ciencias de la Salud, Universidad Europea de Canarias, La Orotava, Tenerife, España

Luis D. Boada
Toxicology Unit, Clinical Sciences Department, Universidad de Las Palmas de Gran Canaria, Centro de Investigación Biomédica en Red de Fisiopatología de la Obesidad y Nutrición (CIBERObN) and Instituto Canario de Investigación del Cáncer (ICIC) Plaza Dr. Pasteur s/n 35016- Las Palmas de Gran Canaria, Spain

Luis Alberto Henríquez-Hernández
Toxicology Unit, Clinical Sciences Department, Universidad de Las Palmas de Gran Canaria, Centro de Investigación Biomédica en Red de Fisiopatología de la Obesidad y Nutrición (CIBERObN) and Instituto Canario de Investigación del Cáncer (ICIC) Plaza Dr. Pasteur s/n 35016- Las Palmas de Gran Canaria, Spain

Patricio Navarro
Urology Service, Complejo Hospitalario Universitario Insular-Materno Infantil, Avda. Marítima del Sur, 35016 - Las Palmas de Gran Canaria, and Instituto Canario de Investigación del Cáncer (ICIC), Spain

Manuel Zumbado
Toxicology Unit, Clinical Sciences Department, Universidad de Las Palmas de Gran Canaria, Centro de Investigación Biomédica en Red de Fisiopatología de la Obesidad y Nutrición (CIBERObN) and Instituto Canario de Investigación del Cáncer (ICIC) Plaza Dr. Pasteur s/n 35016- Las Palmas de Gran Canaria, Spain

María Almeida-González
Toxicology Unit, Clinical Sciences Department, Universidad de Las Palmas de Gran Canaria, Centro de Investigación Biomédica en Red de Fisiopatología de la Obesidad y Nutrición (CIBERObN) and Instituto Canario de Investigación del Cáncer (ICIC) Plaza Dr. Pasteur s/n 35016- Las Palmas de Gran Canaria, Spain

Ana Isabel Dominguez-Bencomo
Grupo de Investigación en Toxicología Clínica, Hospital Universitario de Canarias, Santa Cruz de Tenerife, España, Facultad de Ciencias de la Salud, Universidad Europea de Canarias, La Orotava, Tenerife, España

María Camacho
Toxicology Unit, Clinical Sciences Department, Universidad de Las Palmas de Gran Canaria, Centro de Investigación Biomédica en Red de Fisiopatología de la Obesidad y Nutrición (CIBERObN) and Instituto Canario de Investigación del Cáncer (ICIC) Plaza Dr. Pasteur s/n 35016- Las Palmas de Gran Canaria, Spain

Octavio P. Luzardo
Toxicology Unit, Clinical Sciences Department, Universidad de Las Palmas de Gran Canaria, Centro de Investigación Biomédica en Red de Fisiopatología de la Obesidad y Nutrición (CIBERObN) and Instituto Canario de Investigación del Cáncer (ICIC) Plaza Dr. Pasteur s/n 35016- Las Palmas de Gran Canaria, Spain